HISTORY OF BROADCASTING: RADIO TO TELEVISION

HISTORY OF BROADCASTING: Radio to Television

The Electric Word

The Rise of Radio

PAUL SCHUBERT

ARNO PRESS and THE NEW YORK TIMES

New York • 1971

Reprint Edition 1971 by Arno Press Inc.

Reprinted from a copy in The Newark Public Library

LC# 78-161139
ISBN 0-405-03560-8

HISTORY OF BROADCASTING: RADIO TO TELEVISION
ISBN for complete set: 0-405-03555-1
See last pages of this volume for titles.

Manufactured in the United States of America

THE ELECTRIC WORD

THE MACMILLAN COMPANY
NEW YORK · BOSTON · CHICAGO · DALLAS
ATLANTA · SAN FRANCISCO

MACMILLAN & CO., LIMITED
LONDON · BOMBAY · CALCUTTA
MELBOURNE

THE MACMILLAN CO. OF CANADA, LTD.
TORONTO

THE ELECTRIC WORD

THE RISE OF RADIO

BY PAUL SCHUBERT

THE MACMILLAN COMPANY
NEW YORK MCMXXVIII

Set up and electrotyped.
Published October, 1928.

SET UP BY BROWN BROTHERS LINOTYPERS
PRINTED IN THE UNITED STATES OF AMERICA
BY THE FERRIS PRINTING COMPANY

ACKNOWLEDGMENT

THE author wishes to acknowledge the courtesy and assistance of the Historical Section of the Navy Department, and of numerous officers of the Naval Communication Service, of Captain S. C. Hooper, U. S. Navy, and Mr. W. D. Terrell of the Department of Commerce. He has made free use of factual and statistical information found in H. L. Jome's *Economics of the Radio Industry,* Stephen Davis's *The Law of Radio Communication,* and in the publications of Dr. Irving Langmuir; apart from these he has consulted volumes too numerous to mention here, but to which he is no less indebted. His thanks are likewise due those scientists, engineers and laboratory workers, lawyers and business men of the radio industry, whose interviews contributed much that was not to be found in books, and he is especially grateful to those who read and checked the manuscript.

To this acknowledgment he wishes to add a word of explanation. There is much scientific discussion, these days, as to exactly what is the medium through which this thing, radio, travels. The material "ether," that was once conceived as the substance that bore the waves of Hertz, has gone somewhat out of date, and in its place has been conceived an ether much more flexible—one seen with sufficient imagination not to demand too fine a limitation of its physical characteristics. Some, indeed, would abandon the ether concept altogether, and call the waves "space waves" or "radio waves."

But the word "ether" has been in use for thousands of years. It existed long before anyone knew there was such a thing as radio. It is a most elastic word; it has had dozens of definitions. It has been very closely associated with the radio art since that art's inception, and the author has considered it broad enough to cover "space" in general, and has, therefore, used it throughout the book to express the medium through which radio-communications travel. So, likewise, he has spoken of "ether waves" without any wish to imply restriction of a physics that expands daily.

P. S.

New York City.
September, 1928.

CONTENTS

PART ONE

THE ERA OF MARITIME ADOPTION

PART TWO

THE ERA OF MILITARY USE

PART THREE

THE ERA OF POPULAR USE

PART ONE

THE ERA OF MARITIME ADOPTION

THE ELECTRIC WORD

CHAPTER ONE

A YOUNG MAN GOES TO ENGLAND

THE last decade of that great, forward-striding century, the nineteenth, was two-thirds past when a young man from Italy went up through Europe toward the nation which had set the world's pace during those hundred years. England the mighty, cornerstone of the British Empire and in her own mind custodian of this earth's progress and prosperity—she was a nation that did things. This young man had, he thought, something that would interest her. His name was Guglielmo Marconi, and in that year 1896 he was twenty-two years old.

He was a practical youth. He had as assets culture, education, family, some money but not a fortune, the imagination of his Irish mother, the temperamental realism of his Italian father, and for his chief stock in trade an idea and a kitful of electrical apparatus. He had a mind that was instinctively orderly, organized, and which provided an excellent balance for the abundance of vision that was in it. That mind, two years before, had seized upon a laboratory phenomenon and gone to work to take it beyond the realm of the scholar and bring it to commerce. It thought in concrete, material terms. It repudiated the parental suggestion that in music he would find the ultimate career that a man might have. No, Marconi liked music, but he wanted to leave his mark on men by means of something more tangible.

His idea and his instruments had to do with making practical use of the new "Hertzian æther waves." As a student in the University of Bologna he had seen the "waves" demonstrated by Professor Righi, an enthusiast of science, who had unfolded to his classes those lately discovered mysteries of electricity in space. They had challenged the practical, commercial sense of the young man. They ought, he thought, to be put to work. Their obvious use was the carrying of man's telegraphic messages across spaces which wires could not bridge.

Telegraphy without wires! It was a thing experimenters had been trying to achieve for many years.

So he had gone home for his holidays with the

working out of that problem in his mind. He had bought apparatus, had manufactured what he could not buy. Two things were prominent in his array of material,—one, a conventional telegraph "key," the other, a "Morse Inker"—an instrument for the reception and recording of telegraphic code upon a tape. His purpose had been to actuate the second by means of the first, just as the wire telegraphers did—but he had wanted to omit the wires.

He had succeeded.

More than succeeded. The waves as demonstrated in the Bologna laboratory had traveled but a few feet. If their passage was to be limited to such puny distances, what good would they do the world? It had seemed to Marconi, as he bent his mind against the problem, that if he could give them more of a jumping-off place, they would go farther. So he had launched them from a copper plate held above his work-table by a pair of slim, rickety poles; with another such plate he had gathered them in again at a distance manyfold greater than that of the Bologna laboratory.

Now he was taking it all to England, to introduce it to commerce. Not to make extravagant predictions that he would revolutionize the communications of the world; such romancing he would always leave to others. No, he was merely bearing certain apparatus that he knew would accomplish certain things, so that by its performance, by its concrete satisfying performance, he could demonstrate his achievement beyond question of doubt. Then he wanted to make commercial use of it. If there was one thing in the world that he disliked, it was hocus-pocus. He was not a magician; he was a business man.

In 1896 the Western world—particularly among its northern races—was on the very verge of that complete individual mastery of nature toward which its creative minds had been working since the revival of learning some four centuries earlier. This was the second era of man-knowledge that it had known. The first, two thousand years before, had been the product of Greeks, Romans, Jews. Its contributions had been largely mental and ethical. In mathematics, logic, human relationships, in the government of man by himself, in religion, it had left a profound heritage.

Craftsmen, not scholars, had engendered this second era of knowledge. Craftsmen had devised a printing press, had learned the secret of paper; books, then, could find their way to the common man. Knowledge had come flooding in upon the world with a spread that far surpassed the isolated culture of the early days, and once again man had begun to think creatively.

As if in recognition of the craftsman's part in its making, this time civilization was less concerned with thinking in the abstract than with the material application of new discoveries. The things laid bare by scientists were brought to every man to aid him in his individual conquest of nature. Metallurgy, steam engineering, electricity . . .

The Greeks had known there was such a thing as electricity. In this new revival of learning, men had been fascinated by it more, almost, than by any other phenomenon. It was elusive, hard to classify, impossible to confine. For a century and a half, before 1896, men of this second civilization had been pursuing its secrets intensively. Hundreds of experimenters had set their brains against it, and after 1800, with a rush that was almost, at last, complete mastery, they had begun to lay bare one and another and then dozens of its laws and powers.

In 1800 Volta had discovered the electric battery and provided a new way to make the invisible "fluid" —the old had been by rubbing upon glass, amber or such materials. Hard upon that, the investigations of three men, Oersted, Ampere and Faraday, each building upon the other's work, had led to the first dynamo, and, far more important, to an explanation of its action, so that other men could read and understand, and carry on the work of perfecting it. Then and there, with so much of the ground covered, electrical investigation took two different paths. Some men kept breasting on into new fields; others took up the burden of refining and perfecting the things already learned, making them useful to man.

In 1820 Arago had circled a core of iron with

a current-bearing wire—the electro-magnet. An American, Joseph Henry, had insulated the wire with silk and multiplied the number of turns in his coils into the thousands; his magnets were powerful enough to lift weights of thirty-five hundred pounds. He made tiny ones, too. One such, in 1831, he operated from the other end of more than a mile of wire.

Then came an American artist, a painter of portraits named Samuel Morse, and seized upon this idea of operating electro-magnets at a distance. The telegraph became his passionate quest and he spent five years—1832-1837—in perfecting it. He conceived a "code" by which messages could be transmitted through his instruments. Through his efforts, backed by a New Jersey family of iron-and-brass-workers named Vail, and by an appropriation granted by the Congress of the United States, there was made possible the network of telegraph wires that covered the civilized world in 1896.

In 1896 no other nation was so adequately and efficiently served by the telegraph as was the United States, because no other nation needed or demanded such rapid internal communication. This was a nation whose vast distances were all within one continent, her own territory.

The British Empire, on the other hand, was a union of colonies spread all over the world, separated by countless leagues of sea water, but bound together by common need. Britain, for her communications, had taken the undersea cable, invention of Cyrus Field, another American, and used it to break down the sea barriers. In its development she had demonstrated much the same resource and broad vision that had brought the telegraph to such heights in the United States. In 1896 over 60,000 miles of British cable reached out from London to the Dominions of the Empire and to lands across the seas; this was three-fourths of all the cable in the world at that time.

And now Marconi had a new device for telegraphy. For its full development he, too, needed the encouragement of governments and moneyed men.

There was a man in England—Sir William Preece, Chief Engineer of the Government Telegraphs—whom he particularly wanted to see. Sir William was not only a communications expert and a man able to lend government support; Marconi knew him to be one of those who had been experimenting in wireless telegraphy by the methods called *Conduction* and *Induction.*

Such experiments had been going on for a number of years,—attempts to utilize the earth or water as conductors of the telegraphic current instead of wires; other attempts to pass currents through space from one circuit to another by induction. But wireless communication by either of these methods, though possible across short distances, was unwieldy. Preece's system, for example, had in 1892 been used between the mainland and two small islands three or four miles out in the Bristol Channel. He had stretched great lengths of wire at each point, parallel to each other; impulses given to one of them could be picked up by the others. Unfortunately, the wires had to be approximately equal in length to the distance to be covered; it was more expensive than, and not nearly as practical as running a direct line.

And so when Marconi arrived in England he went to see Preece, and his visit more than justified his voyage. He found a courteous gentleman, able to understand his devices, more than willing to lend support if they should live up to what was claimed for them. The young Italian made preparations to conduct preliminary demonstrations, as Sir William suggested, in rooms of the London Post Office.

His demonstrations were entirely successful—he had known they would be. With Preece's backing he took the apparatus, then, to Salisbury Plain, set it up and began reaching for distance across that great expense. Before the year had turned, his waves had been detected at a distance of nearly two miles; by March of 1897 he had increased this to four, and he was only beginning.

One thing Marconi did soon after his arrival in England, and that was to apply for and receive a patent covering his "wireless telegraphy." "I believe," he said, "that I am the first to discover and use any practical means for effective transmission and intelligible reception of signals produced by artificially formed Hertz oscillations."

This was a statement not destined to be received with any great degree of cordiality in the scientific laboratories of Europe. To some of the eminent scientists who were experimenting with the Hertz waves of which Marconi spoke, his sweeping claims seemed irreverent and audacious, and after a few of these men had visited Salisbury and discovered what was to them a great lack of novelty about his apparatus, and a description of that apparatus had become generally known, the older men reached a state of mind little less than acerbid.

Scientific pique! Had one of their own number arrived at Marconi's results, they might have felt otherwise; had Marconi in a smashing stroke invented something entirely new, they might have felt otherwise; as it was, he was a presumptuous boy. The fact that his work was so frankly commercial made matters worse. They felt that he was prostituting science, demanding money for the achievements of men beside whom he seemed an insignificant cipher; they felt that something ought to be done about it.

Scientists had made it all possible. To begin with, there had been the Englishman, Michael Faraday, back in the early decades of the century. He it had been who established the proof that a relation existed between electro-magnetism and light; he it had been who defined the laws of induction.

Next in the sequence had come James Clerk Maxwell, a Scotsman with a mind so profound that it thought in terms of universes and interstellar spaces. Where other men had regarded electricity as something that came out of a battery, or which, put into an electro-magnet, could lift a weight, Maxwell thought of it as an integral part of all creation. Sifting the known phenomena in that mind of his, he dropped them, one by one, into their allotted places, and in 1873 published a book on Electricity and Magnetism that revolutionized the conception of every worker in the field.

Among other things his book spoke of the "æther"—the word and its definition came from the Greeks; philosophers had used it for centuries in their attempts to explain the universe. But, said Maxwell: "those who imagined æthers in order to explain phenomena could not specify the nature of the motion of these media, and could not prove that the media, as imagined by them, would produce the effects they were meant to explain. The only æther which has survived is that which was invented by Huygens to explain the propagation of light. The evidence for its existence has accumulated as additional phenomena of light and other radiations have been discovered; and the properties of this medium . . . have been found to be precisely those required to explain electromagnetic phenomena."

Maxwell's theory reduced all electric and magnetic phenomena to motions in the form of waves in a material, all-pervading ether. Light, he said, was nothing but an electric phenomenon, a wave in space. Heat was another such. Electricity another. The difference between them lay in the rapidity of their oscillation—a difference in frequency. Light waves were very short, heat waves very long. Between them was a gap, an enormous number of wave-lengths about which almost nothing was known. . . .

The activity which this theory stimulated in the physical laboratories of the world was tremendous. It was a challenge, for though it seemed to explain all the known phenomena of electricity, yet it was only a theory and its scope was so sweeping in concept that there would be necessary hundreds of experiments before it could be definitely accepted as a law, or rejected as an interesting but theoretical piece of pure mathematics—it was years ahead of the experimental progress of its time.

Before anything else could be done, the physicists had to find some means for setting into motion the unknown waves between light and heat, which also meant that they must establish some means of identifying them once they had been launched into the ether. They could identify light with their eyes. How were they to know the others?

It was fourteen years before, in 1887, a brilliant young German Professor of Physics, Heinrich Hertz of the University of Bonn, completed and made known the results of a series of experiments that confirmed Maxwell's theory. Others had learned that electric oscillations of frequencies comparable to those of the theory, could be obtained by combining a Leyden jar and a coil of wire in a circuit. Hertz, desiring to transmit the oscillations to the ether, left a tiny gap in this circuit; as the oscillating current reached this gap, it leaped it in a livid spark —and that leap launched it.

On the other side of the room he suspended a wire ring, and in that, too, he left a gap. The oscillating waves, brushing this wire ring as they passed, induced a sufficient pressure of current in it to cause sparks to jump its gap likewise, and the thing was proved. Thus was Maxwell's theory confirmed; "Hertzian waves" were given to the world.

According to Maxwell, these waves when discovered were to be exactly similar to light waves. They were to travel at the same speed—186,000 miles a second. As a confirmation, Hertz measured their velocity and length, and found that they agreed with the theory. They became, thereafter, definite and known quantities.

Marconi's patent was for the practical use of Hertz waves. He, who had played no part in their discovery, he, three years out of his teens, now offered them to the world of commerce and proposed to patent them! Small wonder the scientists were shocked.

One other thing provoked a storm of controversy, and that was Marconi's use of the "coherer" and "decoherer." This combination of devices was as fundamental a part of Marconi's apparatus as is the vacuum tube of the modern set. It served, in fact, the same purpose. It was his "detector," and it, too, had been the discovery of other men.

In 1879 Hughes had discovered that loose metallic filings would "cohere" when an electric spark was discharged anywhere in the vicinity, and this phenomenon had become an interesting laboratory demonstration. A stick of wood would be coated with powdered copper and included in an electric circuit. Almost no current would flow, since the electricity could not pass through the wood and the loose powder on its surface was a poor conductor. Then a spark would be discharged, and immediately, as the pulverized metal clung together, current would start to flow. The particles would remain clinging together until the stick was given a light tap, when they would separate again, and current would cease to flow.

In 1892 Professor Edouard Branly of the Catholic Institute in Paris, placed the metal filings in a tiny tube of glass, closing them in at each end by metallic plugs to which were fastened the wires of the circuit, and used the resultant device to detect Hertz waves (which had, of course, the same effect upon it as the conventional spark). Oliver Lodge, in England, christened this device "coherer." In this form it was useful only to determine when a Hertz wave commenced, for only the initial "spark" affected it, and it must be tapped to cause the filings to separate again.

Popoff, in Russia, then thought of including the vibrator and hammer of an ordinary electric bell in the circuit through the coherer, so that the instant the filings were drawn together, the hammer—or "tapper"—would strike the tube and cause them to fall apart again. This was called the "decoherer," and it was a splendid advance, for it made it possible to measure the duration of the waves.

None of these scientists had had any other purpose in their work than the increase of knowledge about the Hertz waves. They were seeking devices by means of which the waves could be more accurately detected and measured. When Branly and Popoff heard that Marconi had included the "coherer" and "decoherer" in his patent, and heard that he was using the word "Inventor" in connection with devices that made use of them, they not only stood upon their scientific dignity as injured men, but sincerely felt that they had been victimized. This in spite of the fact that Marconi was using only the idea—in his actual apparatus he had much improved the construction.

The scientific attitude was crystallized when, in July, 1897, a group of some of the wealthiest and strongest men in England joined with Marconi to form the Wireless Telegraph and Signal Company. This company, which was capitalized at £100,000, paid to the inventor £60,000 in shares and £15,000 in cash for the rights to and ownership of his wireless patents, offered its stock to the general public, and commenced operations for the development and exploitation of wireless telegraphy. Marconi became one of the company's six directors and was placed in charge of its development work.

The system of wireless that Marconi took to England in 1896 was so crude as to be little more than laboratory proof that such telegraphy was possible. Indeed, had it rested at that, most of the things said about the youth by the European scientists would have been justified. Now, however, with financial backing and growing support, his apparatus took in rapid succession the strides that brought it substantially to the state in which it—and all commercially practicable wireless—was to remain until 1912.

On the continent parallel developments, spurred on by rivalry resultant from Marconi's claims, went forward in four nations, where men who never quite caught up with the Italian labored to perfect the identical system and each claimed, because of certain minor individualities in his particular apparatus, that his was quite different from, and altogether better than, Marconi's. In Germany, Professor Slaby, with Count von Arco for his assistant, had perhaps the best of these "systems," which was later manufactured by the Allgemeine Electricitäts-Gesellschaft, foremost electrical manufacturer in

Germany. Another German system, the Braun, was sponsored by the Siemens-Halske Company. In Russia, Popoff; in France, Ducretet; and in Italy, Guarini, developed their versions of the art.

Marconi, however, remained the leader, and one great reason for this must be credited to the fact that after their initial period of being, as Oliver Lodge later said, "not unduly enthusiastic," the scientists of Britain recognized the merits of his pertinacious practical work and coöperated with him. Furthermore, through the financial power of his company, he was able to employ expert assistants, to purchase the patents of other men, and to carry on expensive research and development work with greater freedom than any of the others.

The problems immediately before the twenty-three-year-old were those same the overcoming of which has ever since been the goal of men engaged in radio's development—the elimination of "interference" and the breaking down of distance. The immediate answer to the latter seemed to lie in greater power and higher antennae, but for the former Marconi soon found that however well his wireless telegraph would work when exchanging messages between two stations, the addition of a third resulted in a hodge-podge of unintelligible signs on the receiving apparatus. When he attempted to set up two separate pairs of stations he was checkmated by the same "interference" between their signals.

It became evident that if there were any solution to this problem, the way would lie along paths of "syntony," or "tuning." The ether wave was a vibration, an oscillation. Its period was rigid; once launched into the medium it did not change. A wave of certain characteristics could be launched only by apparatus having identical oscillatory characteristics, and had a maximum effect upon receiving apparatus which was yet a third edition of those features.

Marconi thought of the problem in terms of music—perhaps his musical education was one reason why he understood it so clearly. A musical note was a wave of sound in air. It was exactly similar to a radio "note" except that it was in another medium. The note "C," launched from any musical instrument, would still be the note "C" wherever it was heard. It made no difference whether it was launched from a piano, a guitar, a human throat, or a trumpet—but it could only be launched from any one of those instruments if that instrument vibrated exactly to the "C" pitch. And wherever it traveled, if it brushed past another object which would vibrate to that pitch, it set it in motion to some degree. Marconi's problem was to launch his electrical "notes" clearly and strongly, and to have some object that was (1) very vibrant electrically, and (2) that would "vibrate" only at the frequency of the ether "note," to pick them up.

As it was, his "notes" were not at all clear.

His spark-gap—raw generator of waves—was set directly into his antenna. He had no means of striking any one note—it was as though two or three octaves of a piano keyboard were thumped down at one stroke; a great jumble of waves went into the ether, cutting a wide swath and colliding with any other Hertz waves that were there. Nor could his receiving instruments focus on any one portion of the ethereal highway; they were set in motion by every wave-train that came along.

Syntony of oscillating circuits had been one of Oliver Lodge's quests for some time. The laws for making a closed circuit vibrate electrically were known and understood, but in applying them to his wireless Marconi was working in the dark. The antenna was one of his greatest stumbling blocks. Connecting its lower end to earth made, he found, some improvement, but not enough; he saw that he was going to have to divorce it from the source of the oscillations—that instead of connecting the two directly, they would better be connected only by induction. Little by little he realized that for complete syntony five separate elements would have to vibrate in "tune": (1) the initial source of the oscillations—an oscillating circuit containing a spark-gap; (2) the sending antenna circuit; (3) the waves in the ether; (4) the receiving antenna circuit; and (5) the receiving oscillating circuit, containing the coherer and the receiving apparatus.

This realization and its practical achievement

were the labor of two years, and doubly difficult because he had as yet no adequate means of measuring the elusive electrical quantities involved. He tried changing the length of his aerial wire, changing its shape, changing its construction. He came to realize that the antenna proper was nothing but a big "condenser" introduced into the antenna circuit, and to regard it merely as one of the instruments. Not until late in 1899 had he perfected syntonized wireless and applied for patents. His famous patent No. 7,777, issued in April, 1900, was the result of the expenditure of many thousands of dollars, of months, years of highly concentrated labor and experiments with hundreds of different apparatus.

For seventeen years that patent, granted in many countries, was to be the basic radio patent of the world. That patent was the Marconi radio as the world used it. Many refinements were to be added to it, but it was to be years before it could be called obsolete, and its fundamental principles will always apply.

It is safe to say, too, that at the date of the taking out of that patent Guglielmo Marconi saw radio with a clarity approached by no other man on earth. In achievement he was some two years ahead of all the rest; in clarity of understanding he stood alone. And in 1900 he was yet only twenty-six years old.

CHAPTER TWO

MANY SHIPS ACQUIRE VOICES

LONG before syntony was achieved, the new wireless telegraphy was being put to uses indicative of the future that lay ahead of it. Indeed, syntony, although it was the single step that made possible commercial radio on a large scale, was yet a refinement; it permitted the carrying on of simultaneous communications by two or more pairs of stations, but the wireless telegraphy that existed prior to its achievement was infinitely better than none at all.

"Distance" was the key to initial adoption, and happily, the gaining of distance was a thing always far easier than the attainment of resonance. Higher antennae and greater powers sufficed. Even before his new company was formed, Marconi had spanned ten miles across water; within two months after that event signals from his Salisbury apparatus had been recorded at Bath, thirty-four miles away.

With the facilities of the new company at his command, he gave up his experimental station at Salisbury and moved down to the coast. In November, 1897, stations with masts 120 feet high were ready at Alum Bay on the Isle of Wight, and at Bournemouth, fourteen miles away on the mainland. A steamer with a sixty-foot antenna was sent out and on December 6th received his messages from the Isle of Wight station across eighteen miles of water. Work was carried on daily with Bournemouth; in mid-1898 that station was moved to Poole, farther away; a thousand words were transmitted daily in each direction and weather was no obstacle. The gradual improvement of the system made it possible to lower the Poole antenna to eighty feet. On June 3rd Lord Kelvin came into the Wight station and paid money for the transmission of a message to Bournemouth!

More than anything else the ability to send messages across water to and from ships, regardless of fog, weather or light, captured public imagination. Much had already been told of Marconi's activities; in July the *Daily Express*, a Dublin newspaper, brought him new publicity by arranging to have the story and results of the annual Kingstown Regatta sent ashore by "wireless" from the steamer *Flying Huntress*, chartered to follow the yachts as they raced. He set up his apparatus with a seventy-five-foot antenna on board the ship, and a one-hundred-and-ten-foot antenna in the grounds of the Harbor Master at Kingstown; during the three days of the races he sent over seven hundred messages between the two. People were able to read the results of the regatta and gasp as they realized that the yachts had not yet returned to port! The *Flying Huntress* had been ten miles at sea. Before Marconi removed his apparatus he made a few changes and increased that distance to twenty-five.

When, that year, the capital of the company was doubled and another £100,000 of stock placed on the market, the shares were snapped up at a premium. All was going out, almost nothing coming in, yet it was a speculation that appealed strongly to the investor. In spite of warnings by the pessimists, there was growing a profound public faith in Marconi as a miracle man, a faith engendered by the quiet earnestness of his toil and the steady progress of his work. As time went on this strengthened for the paradoxical reason that he was nothing of the sort—he was a cautious, conservative builder.

Little by little the revenue began to come in. In December, 1898, five months after the Kingstown demonstration, the Lighthouse Service authorized the establishment of Marconi radio stations for communication between the South Foreland Lighthouse at Dover, and the East Goodwin Sands lightship, twelve miles away. These stations, operated by Marconi men, worked through all that winter's storms and through them was begun the service in the saving of human life that later proved to be the compelling factor in the maritime adoption of radio. Several times during the winter, warnings of wrecks and vessels in distress were sent inshore from Goodwin Sands, and on April 28th, 1899, the lightship herself was run into by the steamer *R. F. Matthews,* and was able, by her wireless, to call help that saved that ship's company.

Meantime the French government, hearing of the activity in England, Germany and Italy (Marconi had granted the use of his patents to the Italian Navy), had grown interested in the new telegraphy and had asked Marconi for a demonstration. He had erected a station at Boulogne and on March 27th had carried on two-way communication with South Foreland Light at Dover, across the channel, thirty-two miles away. Following this, he placed a demonstration installation on a French gunboat, and, with shore stations available at either end of the crossing, the Channel steamer *Princess Clementine* was equipped with his apparatus.

In that summer of 1899 there was another demonstration that meant much to Marconi's wireless. He had always had the friendliest relations with the British government; one of the reasons for the choice of the Isle of Wight site had been its proximity to the naval base at Portsmouth. In that summer Marconi apparatus was placed aboard three of Her Majesty's battleships and used under service conditions during the naval manœuvres. Distances up to eighty-five miles were covered, and the British Navy immediately purchased a "complete outfit" and installed it in the *Defiance,* torpedo school-ship at Devonport, for further experiments by naval officers.

By this time it was definitely established that the commercial use of wireless was, for the time being at least, to be chiefly for the maintenance of communications between ships and shore, and between ships and other ships. Gradually two different types of radio instruments were taking shape, one particularly adapted to shore use, having a great amount of power and high masts, the other adapted for ships, where both the height of antenna and the power were limited. Although the much desired syntony had not been quite achieved, the work thus far had produced refinements that made distances of one hundred miles possible and distances up to one half that certain under all conditions except through interference.

So much had been accomplished when that enterprising newspaper, the *New York Herald,* engaged Marconi to bring his instruments to the United States for the purpose of sending "wireless" reports of the International Yacht Races to be held in October, 1899, and as a result the new art crossed the Atlantic and made its American début.

Marconi and his company had been thinking about the United States, as they had been thinking about all the nations of earth. Ultimately they wanted stations in America; they hoped, too, to interest the United States Navy, just then rather prominent in Europe because of its decisive Spanish War successes and because of America's recent embarkation upon a building program destined to bring its naval strength nearer that which a first-rate power should have.

Accordingly, it was decided to take advantage of the trip to the States to accomplish other things than newspaper telegraphy. It was decided to demonstrate the apparatus to the United States Navy in the hope of interesting that service in its adoption on a large scale, and for purposes of following up this and other commercial exploitation of the "wireless," it was decided to form an American company. Business representatives of the British company accompanied Marconi on the trip; they were to arrange for the financing and incorporation of Marconi's Wireless Telegraph Company of America.

The American yacht *Columbia* beat Sir Thomas Lipton's *Shamrock* on October 16th, again on the 17th and for a third time on the 20th; Marconi's wireless carried the news ashore and that fact was featured in the *Herald*. The races over, the Italian proceeded to carry out the Navy demonstration. He set up his apparatus in the battleships *Massachusetts* and *New York*, and ashore on the Highlands of Navesink on the New Jersey coast, just south of New York harbor.

The first tests were at short distances while the two ships were at anchor in the Hudson River. *Massachusetts* then stood to sea, and recorded *New York's* signals until she was past Sandy Hook, though difficulties in the latter ship prevented replies from being received. Finally, *New York* moved down to Navesink and anchored, *Massachusetts* proceeded to sea for distance tests, and the shore station on the Highlands played its part by acting as the source of the "interference" about which the naval officers had heard so much.

When the Board of naval officers made their report they said: ". . . the Marconi system of wireless telegraphy . . . is well adapted for use in squadron signaling under conditions of rain, fog, darkness, and motion of speed. Wind, rain, fog, and other conditions do not affect the transmission through space. . . . Darkness has no effect. . . . The accuracy is good within working ranges. . . . The greatest distance that messages were exchanged with the station at Navesink was 16.5 miles. This distance was exceeded considerably during the yacht races. . . ."

Interference, said the report, had been complete when attempted, and though Mr. Marconi had stated to the Board, before the tests, that he could prevent it, he had not explained how, nor had he made any attempt to demonstrate that this could be done. It went on to caution naïvely that the shock from the sending coil might be dangerous to a person with a weak heart, and that the sending apparatus would affect the compass if placed near it; it summarized the uses to which the "wireless" might be put, and closed with the respectful recommendation "that the system be given a trial in the Navy."

The tone of the report was not as enthusiastic as it might have been, but the demonstration had not been nearly as effective as that in the British Navy. Marconi wrote a letter explaining that he had been unable to demonstrate the devices used for preventing interference by tuning because they had not yet been completely patented and protected . . . he had explained before the tests that the distance would be less than that obtained in the British naval manœuvres, as the apparatus was merely that brought for the yacht races where long-distance transmission was unnecessary.

On November 22nd, 1899, the Marconi Wireless Telegraph Company of America was organized under the laws of the state of New Jersey. Its authorized capital was $10,000,000, covered by the issuance of two million shares of stock with a par value of $5.00 each, and most of those shares took their way back to England as the property of the British Company, for the Marconi Wireless Telegraph Company of America was the first of the many subsidiaries which that company was to form. Its purpose was the carrying out of the American end of the operations of the British company.

Marconi had not waited for this event. After

completing the naval demonstrations he had sailed for England in the *SS. St. Paul,* and had as usual seized the opportunity to experiment. As the steamer neared the end of her crossing he was listening for his stations on the Channel. On the fifteenth of November he picked up the Isle of Wight, thirty-six miles away; there was news from South Africa and the Boer War, and the ship's officers incorporated it into a leaflet newspaper designated *The Transatlantic Times.*

With the turn of the century there came also a turn in the affairs of radio; 1900 was the first year of the movement toward its general adoption aboard seafaring vessels. The British Navy had followed up its initial purchase by ordering five more sets which were in use on ships in South African waters; in May it contracted for apparatus for twenty-six ships and six coastal stations. In February demonstrations were made in Germany; permanent stations were installed on Borkum Island and Borkum Light-ship, and the North German Lloyd liner *Kaiser Wilhelm der Grosse* was equipped. Steamship companies, both British and German, began to show considerable interest; foreign navies were ready to follow Britain's suit.

This activity made necessary the adoption of definite commercial policies by the young company. The fundamental principle of their business was, they decided, the sale of communications; they were a telegraph company. Their plan of operation was as follows: (1) they would maintain adequate shore stations to serve the vessels equipped with their apparatus; (2) they would not sell shipboard apparatus, but would lease it and provide a capable operator with it, at an annual rental that would cover royalties, operator's salary, and the maintenance and depreciation of the apparatus, the operator being an employee of the Marconi Company; (3) over the telegraph system thus established they would transmit commercial "Marconigrams" (to and from the ship's passengers) at rates to be established, and would handle all messages between the captain, owners, agents, etc., at much lower rates.

To carry out this plan the company was reorganized in March, and its name changed to Marconi's Wireless Telegraph Company, Limited. This was to be the parent company of the system, and its direct concern was the erection and operation of the necessary British shore stations, and the establishment of foreign subsidiaries, such as the already created American Marconi Company, for the same purpose in other lands. For the operation of the ship-rental service, a new company, called the Marconi International Marine Communication Company, also subsidiary to the parent company, was formed.

So organized, the company pushed ahead vigorously, for already continental competition was making itself felt. By the end of the year it had, in addition to the half dozen shore stations already in existence, put up its now familiar masts at such widely separated spots as Fastnet Rock, at the western end of Ireland, from which point it sent reports to the Board of Trade of all incoming steamers sighted; on Nantucket Light-ship, south of the Massachusetts coast, where, sponsored again by the *New York Herald,* it sent similar reports to that newspaper; and at Lapanne, in Belgium. A Belgian Marconi company was organized in 1901.

These stations were erected as part of a commercial telegraph system. Pursuing an opposite policy, the European manufacturers of wireless equipment (who, in competition with the Marconi Company, were trying to interest ship-owners in buying—not leasing—their sets) did not evidence any great activity in erecting competing shore stations, and the Marconi Company, spending thousands on such stations, now began to be called on for service by ships equipped with rival apparatus. This service they decided not to render—rules were established that only in case of an emergency would a Marconi shore station accept a message from other than a Marconi-equipped vessel. This was a stroke that told heavily on the competing companies, particularly those in Germany.

The rental idea worked out very well for ships of the merchant marine, but when it came to navies

—a field that it was also desired to cultivate—it was obvious that it would not be acceptable. Foreign navies wanted to do as the British Navy had done—buy their apparatus outright and operate it themselves. The Marconi Company, though reluctant to accede to this, realized that it was inevitable—it was impossible to expect the captain of a man-o'-war to look for his wireless to a Marconi operator who was not even a member of his own crew. But the company felt that it should be compensated for its expensive development work, and further, that sales to navies should be prevented from serving as stepping-stones for the advancement of rival companies or the facilitation of their entry into the radio-communications field.

Accordingly, when the navies to whom Marconi wireless had been demonstrated asked to be quoted prices, they were told that the Marconi Company would equip them on the following terms: the navy in question was to sign a contract to use Marconi wireless for a long term of years; it was to pay a flat royalty of £10,000 a year; it was to purchase its instruments at the current market price, including duty; it was to agree to accept messages for relay from merchant vessels equipped with the Marconi system; it was to undertake to prevent interference between naval vessels and commercial Marconi stations; it was to agree not to use Marconi wireless for communicating with any other system of wireless telegraphy except in an emergency or when communicating with another naval vessel. In return, it was to acquire certain privileges in the way of rates, time, etc., at all Marconi stations, which would make it unnecessary to build as many shore stations as it would otherwise have to do.

The Marconi Company felt that nothing less than these terms would be fair to itself, both because of its past work and of the future development which it was shaping. They were received by the continental governments, however, with a certain sense of shock, and capital was instantly made of them by the European competitors.

Marconi's later work had all been kept as secret as possible. None of the new "syntonized" outfits —now nearly perfected—had been released, and apart from the knowledge that he was at work on syntony, no one on the continent had any idea how far he had come or of the records which he was to set in 1901. On the other hand, Germans, Frenchmen, Russians, all had by the end of 1900 apparatus comparable to his non-syntonized instruments. He could work fifty or sixty miles, a hundred occasionally; most of them had done likewise. They told their respective governments that the Marconi royalty demands were nothing short of piracy, and hinted at a dark British desire to "listen-in" to the radio talk of the continental navies. Marconi wireless, they said, was not secret; they had, themselves, "listened-in" to British ships working—copied down every word. But each of them—and they had all made some progress along syntonic lines—claimed to his respective government that *he* had apparatus with which no one not similarly equipped could communicate.

Rejecting Marconi, then, the French Navy equipped with Popoff-Ducretet apparatus, Germany resolved in favor of Slaby-Arco, Russia (which had a navy larger than that of the United States) ordered two hundred Popoff sets, and Sweden, after trying the Siemens-Halske Braun system, followed Germany to Slaby-Arco. So early, radio began to take national lines.

In the United States, where almost no radio experimentation had been carried on and where the Navy objected to the Marconi terms just as much as they had been objected to in Europe, the matter was, for the time being, dropped.

While the business members of the Marconi Company were engaged in these matters of expansion, Marconi himself had syntony within his grasp, had acquired all the patents necessary to it, and was very busy working out a set of instruments which would best embody his new discoveries. He had, he knew, something infinitely better than anything previous. The wine of success was in him. He drove about the country in a steam-automobile which he had equipped with his apparatus, using a large sheet-metal "condenser" for an antenna, and when, as sometimes happened, he had a visitor with him watching his tests, he amused that friend by sending back messages to his hotel, saying that they would return at such and such a time,—would the hotel please have the following dinner ready. . . .

To return and find it waiting, just served and piping hot—that was alchemy. Men's imaginations, then, would stray, and they would wonder where it all would end.

When 1901 dawned he was ready for things beside which all his former work would seem trifling. As a preliminary test of the new equipment he set up a station at the Lizard, far down on the southwest tip of England, 186 miles from the Isle of Wight. The first attempt at communication between the two set his coherer tappers clicking and the tape unwinding from the receivers, bearing upon it the clear code marks of success. . . .

It was enough. He decided to build such a station as had never been built before. He decided that the time had come to make a real assault upon distance. In a day when fifty was considered good, 186 miles was startling, but with what he knew about the new instruments, it was not startling to Marconi.

At Poldhu in Cornwall—that same southwest part of England—he set his men to work erecting a station. It was to have twenty masts, each two hundred feet tall, arranged in a circle two hundred feet in diameter. Nestling beneath these towering spars was to be a cluster of buildings, tiny by comparison, housing his most powerful transmitter to date. The station was to serve for telegraphing to ships far at sea—that, he knew. He believed it *might* serve to communicate with the Western continent, North America—that, he intended to find out.

In August, when the station had taken shape (with the masts up, the many long antenna wires streaming down and the apparatus nearly completed) there was a distressing delay; a gale, blowing up, wrecked those tall spars that had been so slowly completed. Perhaps, though Marconi did not know it then, it was just as well, for men learned later that, as compared to summer, winter was a far better time for long-distance wireless work. At any rate, tests of the new station were deferred until it could be rebuilt, and rebuilding was a difficult proposition. It would be foolhardy to duplicate the first effort—another gale might wreck it again.

Ten masts were re-erected, shortened to 170 feet and heavily guyed, and the entire aerial design was changed; a "fan" of sixty wires, each a yard apart at the top, converging to a point at the bottom, replaced the "gas-tank" effect. It was all ready, finally, in mid-November; the strength and clarity with which its signals were received at Crookhaven, Ireland, when it was tested, were so great that Marconi decided to sail for Newfoundland immediately—he believed he would be able to receive them there, on the other side of the Atlantic.

Like all his comings and goings, this one was very quiet. Marconi never promised; he spoke only to tell of achievement. Something of an air of mystery always surrounded him, engendered by his reticence. With two assistants, the necessary receiving apparatus, and a number of kites and balloons to support the test antenna, he took ship in the *SS. Sardinian* on the 26th of November, and arrived at St. Johns, Newfoundland, ten days later. In making his preliminary arrangements he let it be generally assumed that he was going to work with ships off the coast—only a few knew the secret. He did not want it said, if he should fail, that he had had a foolish dream that his waves would or could cross so great a distance. He loved stability, hated to be thought of as a visionary.

On December 9th he had selected Signal Hill, overlooking the harbor, from which to fly his kites, and set up his instruments in a room in a hospital there which had once been a barracks. He wanted as high an antenna as he could get; the next day he sent a kite up six hundred feet, and on the day after that a balloon, but the balloon broke away. He decided on the kite for the actual test, and cabled Poldhu to begin sending on the following afternoon at 3:00, and for three hours to repeat the prearranged signal—the letter "S," three dots in the Morse alphabet. The equivalent time at St. Johns would be 11:30 A.M., to 2:30 P.M.

The morning was raw and gusty, with fog swirling in off the Newfoundland Banks. Marconi, bundled up in an ulster, stood by while his men struggled with the big kite. They got it up, up four hundred feet but no higher. Then, the hour come, he went inside the building to his instruments.

For the test he had abandoned the "inker"—it was not sensitive enough. In its place he had connected a head telephone; he put it on and stood waiting, wondering whether he would hear anything, scarcely daring to hope. Many men had said the waves would not follow the earth's curve for such a distance, that they would go off on a tangent into space; there was no way of telling the truth of that except by hearing or failing to hear them. His assistants were in another room; the only sound was the roar of the sea below and the roaring of his straining ears.

He waited so for an hour. Then, at half past twelve, there was a sound in the telephones that he could scarcely believe, a sound that set his heart to beating wildly and his hand to trembling. For a moment he wondered whether his tensed senses had not hypnotized him into believing that he heard that longed-for letter "S." But his instruments had come to life. The tapper on the coherer was unmistakably clicking, the telephone was breathing its faint rustle, and with a feeling of sweeping exultation he exclaimed: "Gentlemen, did you hear it!"

They came running, and he handed the instrument to one of them, pleading: "Can you hear anything?"

Yes, beyond question the Western Ocean had been spanned. The signals were barely audible, but there was no denying that they were being received. Delighted, ecstatic, they passed the phones from one to another. .. . .

When, next day, the test was repeated, depression had overtaken Marconi and driven that first high exaltation from him. He shrank, now, from telling the world. So many things had been said of him; he knew this would bring fresh accusations from certain quarters. It was, after all, his word against theirs. But, he thought, what did that matter so long as he knew, himself? No, it was the realization of how much there was yet to be done, how far there was to go, that rested heavy upon him. Yesterday that letter 'S" had seemed miraculous; today it was, already, nothing. Nothing would count, now, until he was sending entire messages back and forth, every day, as well as the cables did.

One thing *was* possible. The Governor of Newfoundland was one of those who had been told exactly what was going on—he would arrange a demonstration for that official's benefit, and then there would be disinterested witnesses who could verify his statements. . . .

He announced his achievement, then, and the plaudits began to come in, but he never gave the demonstration for the Governor, for an episode intervened that must always remain a choice bit of ironic humor. The St. Johns officials of the Anglo-American Telegraph Company—the cable company—notified Marconi in all seriousness that since they had a charter giving them the exclusive right to erect and operate stations for telegraphic communication between Newfoundland and places outside the colony, he was engaged in work that violated their rights. He, said they, must stop all work immediately and remove the apparatus or legal proceedings would be taken against him.

Marconi was stunned. To have come so far, to have done so much, only to be halted for such a reason. . . . But, as he complied with the order and packed away his instruments, his spirits, which had been so low, revived in a soaring leap. He chuckled. They took him seriously, did they? They took his wireless seriously? Well, he'd show them that they had reason to.

He wanted, now, to erect a station on the North American continent that could return communications to Poldhu. All his interest was in this new aspect of the work—transoceanic communication. For all he knew it might be possible even now to operate regularly, commercially, and he burned to try it. Before he returned to England he had been flooded with offers of land for a site; he chose that of the Canadian government for a location in Nova Scotia, and went home to confer with the company's directors.

He stayed in England less than a month, for it seemed to them all that this was a chance, at last, to break into real revenue, and that the Canadian station must go up as soon as possible. Marconi was aboard ship again on February 22nd, 1902, bound for Nova Scotia via New York, in the American Line *SS. Philadelphia.* For 1551 miles of the voyage he received messages from Poldhu; until the ship was 3000 miles out he was still receiving single letters. Already, after a mere three months, the St. Johns tests seemed puny.

The Marconi Wireless Telegraph Company of Canada was formed to handle the finances of the new venture, and in March contracts were signed for the erection of a station, equal to Poldhu in size, at Glace Bay, Cape Breton, Nova Scotia.

It was nine months in building. December had come before its waves went out, but on the seventeenth of that month Poldhu heard its messages and the Atlantic had been crossed for the first time from the West. On the eighteenth the formal messages that mark such an opening were exchanged; the Governor General of Canada and his King, in England, addressed each other by wireless.

The American company, too, built a station for the new transoceanic communication, and opened it just a month after that in Glace Bay. On January 19th, 1903, President Roosevelt's greetings to Britain's sovereign likewise traveled through space. . . .

For the sake of drama and dreams it would be satisfying to be able to say that thereafter radio took its place as the best and cheapest means of telegraphing across the seas, but such, unhappily, was not the case. Both these stations were disappointments, for though transoceanic communication was possible, regular service was not. Only under the most favorable conditions could messages be exchanged, and young Marconi realized regretfully that his dream of world-wide wireless as a worthy rival of the cable was still far from realization.

CHAPTER THREE

BEGINNINGS OF A NEW RADIO IN A NEW LAND

So far, the only commercial radio in the United States had been Marconi wireless. In the other two branches of the art—government use and experimental development—almost nothing had been done. The Navy had had its 1899 demonstration and decided against the Marconi system; Captain Samuel Reber of the Army Signal Corps had attended the demonstration, and when the Army, like the Navy, rejected the Marconi proposals because of their restrictive clauses, that officer had built two home-made sets of instruments for Army use, and with them communicated between Governor's Island and Fort Wadsworth in New York harbor; in 1900 the Weather Bureau had begun experiments at Cobb Island, in Hatteras Sound, thinking to use the new telegraphy to supplement the wire system by which it gathered and disseminated weather information along the coast. Private experimentation, in spite of the activity in Europe, had been confined to the work of two or three men—the nation which had brought the telegraph and telephone to such perfection had scarcely lifted a finger toward the new medium of communication.

Characteristically, however, the little work that had been done had shown strong tendencies to branch off in new directions, and with the stimulus that was now given to the art by the transoceanic activities of Marconi, two American radio companies were formed whose activities, though commercially futile, sowed seeds that later revolutionized the entire conception of "wireless."

The foremost of these was called the National Electric Signalling Company, and was built around the development work of Professor Reginald A. Fessenden, who had been carrying on the Weather Bureau's experiments. Professor Fessenden was Canadian-born, and had been educated in Bishops College, Quebec, and Trinity College, Ontario; he had done electrical engineering for Edison and for the Westinghouse Company, and had been for a time Chief Chemist in the Edison Laboratory at Orange, New Jersey. In 1893 he had become Professor of Electrical Engineering in the Western University of Pennsylvania, at Pittsburgh, and his experiments with Hertz waves had begun in the laboratories there as early as 1895. Four years later he had built a set of wireless instruments on the Marconi idea, and given a demonstration in Pittsburgh; his connection with the Weather Bureau had followed.

Fessenden had been interested in other applications of the Hertz waves than Marconi telegraphy. Radiotelephony, then but little investigated, had been an alluring possibility; it grew to be his chief quest.

Marconi's waves were not suitable for telephony. They were "damped," irregular; each group of them started with a high surge and then died away, so that a Marconi signal was like a ringing doorbell—a train of rapidly repeated single notes.

The sound waves of the human voice were irregular, too. Fessenden knew that if they were converted into electrical vibrations and superimposed upon a Marconi wave, the result would be quite unintelligible. No, he must find some way of generating "undamped" or "continuous" radio waves—waves the crest of each of which was exactly as "high" as those before and after it; such a wave in sound would be that given out by a violin played on one note. He could, he knew, use "continuous" waves as carriers of the voice.

In 1892 Professor Elihu Thomson, the eminent American electrician, had discovered that a direct-current "arc"—the bow of flame between two electrodes connected to a powerful source of electricity—could be used to produce "undamped" oscillations in a closed circuit; Fessenden applied this discovery to radio, and launched "undamped" waves into space. In this apparatus he then included a microphone, and, speaking into it, superimposed his voice upon the radio waves. This was in 1901.

When it came to receiving the new waves he was likewise in unexplored territory. The coherer would not do it; it responded only to the dots and dashes of "damped" waves. There was need, now, for a new kind of "detector," something altogether different from the coherer,—specifically, something which, when the incoming oscillations reached it, would allow the passage of only that half of the current travelling in one direction and would stop the other, oppositely-bound, half. This quality of eliminating the "reverse flow" from an alternating current, thereby changing it into one the entire effect of which was in one direction, was known as "rectifying" the current. A rectifying detector would enable him to receive his telephony.

He found it, at last, in 1902—a combination of electricity and chemistry. Certain chemical solutions would allow a current to flow through them only in one direction. He made a tiny aluminum cup, then, to contain the fluid, dipped a fine wire into it as one electrode, and christened the device a "liquid barretter." Included in his receiving circuit, it was his detector. A second great break away from Marconi had been made.

Now he proposed a third. Why, he thought, use all these arcs and spark-gaps to launch the radio waves? Why not simply connect a source of alternating electric current— an "alternator"—to the antenna?

The answer, for the time being, lay in the difference between the rapidity of oscillation of the current in an alternating current generator, and that of Hertz waves. Alternators like those used to generate electric light current, changed the direction of their current-flow 120 times a second; Hertz waves made the switch from twenty thousand to sixty million times a second. An alternator, then, would have to be speeded up at least a thousand times as fast as such a machine had ever been made to run, if its oscillations, launched into an antenna, were to make radio waves. When Fessenden spoke to other engineers of such a possibility many of them laughed at him. He was told that the whole idea was ridiculous, that no machine could ever be built to rotate so fast, and that no matter how fast the machine was speeded up it could never function as a radio transmitter without a spark-gap or an arc of some sort. Nevertheless he persisted in his belief (which was later to be materialized and to play a most important part in radio) and went so far as to draw up plans and take out a patent.

The development of all these ideas was beyond the scope of the Weather Bureau's activities; that department had neither money nor need for such unorthodox dreams. But early in 1902 two Pittsburgh bankers, Messrs. T. H. Given and Hay Walker, Jr., who knew Professor Fessenden, decided that in the light of Marconi's successes there was a future in transoceanic radio-communications, and that they would back the Professor until he had perfected his telephone system and his alternator.

They formed the National Electric Signalling Company, capitalized for $100,000—the two bankers contributing the cash and Fessenden his patents and services. Given, now, an entirely free hand, he moved up to Chesapeake Bay and built stations at Old Point Comfort, Cape Charles and Ocean View to experiment to his heart's content.

There was another radio engineer whose name was beginning to be known in the United States—Lee De Forest, a young man, no older than Marconi. De Forest had grown up in Council Bluffs, Iowa; had come east and worked his way through Sheffield Scientific School in Yale University and gone on to do post-graduate work there, specializing in researches in Hertz wave phenomena; in 1899 he had gotten the degree of Doctor of Philosophy.

He was a restless, inventive type. After leaving Yale he had gone to Chicago and continued his studies for a time in the Armour Institute; in 1900 he had been in the Test Department of the Western Electric Company at $8.00 a week; in 1901 he had built and demonstrated a wireless apparatus following Marconi's ideas but using innovations of his own.

De Forest was the first young disciple of the new era of American electrical engineering to enter the radio ranks. He had, too, a streak of sheer creative ability that served radio well, though he had, perhaps, less of the painstaking patience necessary for the development of an idea to its utmost than had Marconi or Fessenden, if such dissimilar types may be compared.

What he did in that first radio development of his was to make a conventional "spark" set without the complicated mechanical features that were so difficult to adjust in all the European apparatus. Marconi got his source of power from induction coils; De Forest used a small generator and a transformer—heavier, but less complicated. Marconi used the coherer and Morse inker; De Forest adopted the liquid barretter, made it in his own form with his own chemicals, and called it a "responder."

He, too, wanted a career in the new art. In 1901, after his demonstration, he went to New York to seek backing. For awhile his quest was discouraging, but late in the year he interested a group of Wall Street men who organized the American De Forest Wireless Telegraph Company, which was to manufacture radio apparatus for whomsoever would buy, and to engage in the radio-communications business in competition with Marconi; it was to be financed by sale of stock to the general public.

The winter of 1901-02 was a time of large hopes in radio. Even Marconi, pushing through his plans for those first transoceanic stations of his, thought the art on the verge of sweeping successes. The

American De Forest Company, spurred on by sales of De Forest apparatus to the Army—who thought it the best they had tried—laid out a gigantic wireless scheme on paper and offered it to the world.

"Wireless telegraphy will make of the Pacific an American lake," said De Forest. "The Caribbean Sea will become a telephone exchange." They were going to erect a station at Montauk Point, Long Island, to communicate with England; a complete system in the Caribbean; a complete system across the Pacific from San Francisco to China by way of Hawaii, Guam and the Philippines. For the British end of things they formed the De Forest Wireless Telegraph Syndicate and proposed to erect a dozen stations in Britain. . . .

These large plans they brought before the public with all the display possible, for none of the program could materialize unless stock could be sold to finance it. A "latest model" 1902 automobile ran about the streets of New York carrying a demonstration De Forest apparatus; its spark gap crackled daily before gaping crowds, and every afternoon it invaded Wall Street and stopped before the Stock Exchange to telegraph the "closing prices" to mythical listeners. At Coney Island, the city's amusement resort, a high mast went up and there, too, hundreds were introduced to the new telegraphy.

But the time was not yet. Unfortunately for dreams and promotions, neither De Forest's nor anyone else's apparatus could bridge the oceans in

1902. The grandiose platform and the gullibility of a public that took credence in it constitute an amusing episode of the early days of the art—though it was not so amusing to the many people who lost money in wireless speculation. Lack of capital, lack of a balanced idea of the time's radio-communication possibilities, and lack of willingness to begin modestly and grow, inevitably brought these promoters to grief. They got into financial difficulties and the De Forest Company lay dormant as far as communications went.

It continued, however, to do some manufacturing for the government and others. De Forest himself, turning, like Fessenden, to new fields, grew interested in telephony and commenced fresh experiments in that line.

Coincidentally with the forming of these early American manufacturing and development companies, the Navy took its first toddling steps in a radio direction.

The United States Navy had been a woefully inadequate service for many years. Following the Civil War it had been allowed to lapse into such desuetude that it had hardly been worthy of its name; a revival of interest in the late eighties and early nineties had made possible its Spanish War triumphs, but those had been achieved by a still puny force—four battleships and a handful of lesser craft. Now, due to a general heightening of national world-consciousness, it was at last being built up to something like adequacy. In 1900 ten new battleships were building.

Within its ranks, likewise, there was a great revival of activity, as officers and men found themselves called upon to catch up with the latest developments in their profession and the times. Among these was the use of radio. The Navy had temporarily rejected radio after Marconi's demonstration in 1899, but now intelligence reports from the continent brought the news that every European navy had adopted the art,—said, too, that in Europe apparatus had been perfected equal to Marconi's (for there was little way of judging except by the claims of the manufacturers).

There being as yet no American companies (this was in 1901) the Navy did the only thing possible and turned to Europe. In that year a Lieutenant and two Chief Electricians were sent abroad to study the various systems and bring back two sets each of the apparatus considered important. These men returned in 1902, bringing Slaby-Arco, Braun, Rochefort and Ducretet instruments for tests.

Then began a year's labor. The delicate coherers, tappers, relays and Morse inkers were a very different proposition to American naval officers and enlisted men than to the experienced Europeans who had devised the instruments—much of the time it was a matter of making the "wireless" work at all, rather than of striving for distance, and even while

Marconi was fretting because he could not telegraph across three thousand miles of ocean with regularity, the Navy was exceedingly happy to get ten.

One party operated in a small scale house near the Commandant's quarters in the Washington Navy Yard, where they shared space with a wooden-legged veteran who did the weighing and hated the "sparks" their instruments made; the other end of their "system" was in the Physics Building of the Naval Academy at Annapolis. Later in the year *U.S.S. Prairie* and *Topeka* were assigned to work with them as test ships, and daily steamed up and down the Chesapeake, getting the signals at a greater or lesser distance, or not at all, while the pioneers delved into the mysteries of antennae, induction coils and spark gaps.

In the spring of 1903 the Slaby-Arco apparatus—with which the German Navy was equipped—was finally decided upon as the best of the four systems being tested, and a dozen sets were ordered from Germany. Late that summer these arrived and were installed in the battleships of the North Atlantic Fleet, which was growing to be a respectable force. Signalmen and electricians were trained in their use by the men who had carried out the tests, and when, in the autumn manœuvres, the Fleet divided and fought a sham battle in Long Island Sound, the "Blues" beat the "Reds" through the aid of a radio message that rallied the forces in the dark.

Approval of wireless was not yet quite unanimous; there were Admirals and Captains who were unalterably opposed to it, who believed that when a ship was out of sight of land she belonged in the hands of her master and that orders from the blue were an outrage and an affront to dignity! But there were others who became devoted to the art, and slowly and steadily it progressed in its new utilization. Six shore stations were started that year, and finished soon afterwards. Experimentation went ahead, too; improved foreign apparatus was tested, and also that of De Forest and Fessenden. The Bureau of Equipment, in the Navy Department at Washington, established a special Radio Division; a school for the training of operators was started in the New York Navy Yard.

By the end of 1904 the sea service had become the most ambitious American user of the art it had been so slow to adopt. Twenty shore stations were operating, ten more building, and another fifty proposed! Twenty-four ships were equipped, and sets had been ordered for ten others; sixty-eight were to follow when the best equipment had been finally decided upon! The Navy had been late to start, but once in, it was in to stay. It was, by the way, a new Navy. The number of ships it was outfitting was indicative of how it was growing.

In 1904 there were, even apart from those of the British Navy, some fifty wireless stations in England, some of them commercial, some experimental, some amateur—many of them interfering with each other's operations. Moreover, German, French and American companies were announcing plans to build stations there in competition with Marconi—a circumstance not altogether to British liking. The question of controlling the situation was brought up in Parliament, and on August 15th that body passed the Wireless Telegraph Act of Great Britain, which placed the Post Office in charge of radio. No station could be erected, then, unless it had a license, and any station that abused its license could be shut down.

There was interference in the United States, too, in 1904, and with the growing use of wireless there were several minor squabbles between the different government departments that operated stations, and also between those departments and the commercial radio companies. To iron out these difficulties, President Roosevelt appointed a Board of four men—representing the Navy, the Army, the Department of Commerce and Labor, and the Department of Agriculture—to study the situation and make recommendations.

In August this Board reported. The Army, it said, should be given charge of all government radio in the interior; the Navy, of that on the coasts. The Weather Bureau should get out of the game and let the Navy take over its radio work. The Navy, continued the report, was not to depend upon the commercial companies for service; it was to erect its own system of shore stations for operation in peace and war, the best in the world, if possible. The Navy was, too, to build high-power stations in the Philippines, Guam, Hawaii, Cuba, Porto Rico and the Canal Zone, whether there were private stations in those places or not, so that the Government would have its own communications with our insular possessions.

Commercial operations should, it said, be regulated under the supervision of the Department of Commerce and Labor.

The report of this Roosevelt Board, though it never became law, influenced the whole future of radio in the United States—if not in the world—by committing the Navy to a giant system of its own. It paved the way for government expenditure of hundreds of thousands, millions, of dollars, in the attempt to make that system "the best in the world," and so doing, provided for the future existence of half a dozen American manufacturers of radio apparatus who would otherwise have had no market for their products.

Unwittingly, too, it made rivals of the Navy and the American Marconi Company, for the Navy, as its system grew, came to be a fervent advocate of government ownership of all radio in the United States.

CHAPTER FOUR

RADIO BECOMES A MARITIME NECESSITY, BUT THE ATLANTIC REMAINS AS WIDE AS EVER

BETWEEN 1904 and 1912, radio's inventors and developers worked slowly onward toward those will-o'-the-wisps, telephony and regular transoceanic service. Commercially, use of the art at sea progressed during that period until it had ceased to be a maritime luxury and become a necessity. Strangely enough, the very quality of Hertz waves that Marconi and many of the others would have liked to overcome—the fact that they radiated equally well in all directions—proved to be the fundamental reason for their widespread maritime adoption. They were useful for point to point telegraphy, but vitally essential for broadcasting distress signals from vessels in peril, and on the latter ground their use spread to merchantmen on the seven seas—

many a ship that would cheerfully have done without a telegraph office, equipped with radio for the sake of safety.

With growing adoption came the need for uniformity of operation, and for international agreement on some vexing problems.

In 1903 Germany had invited all interested nations to send representatives to a Radiotelegraphic Conference to be held in Berlin on August 4th, to establish guiding rules for radio. The Conference had met with all the great powers and many of the lesser ones represented, and had readily agreed upon a solution to one previous source of difficulty—the division of wireless tolls between ship stations, shore stations and land-wire companies, the three agencies involved in the handling of messages. There had been no such harmony, however, when the cat came out of the bag and Germany introduced an old plaint. She had proposed that coastal stations be required to accept messages regardless of the system of the sending ship, and that all nations make public such details of their wireless apparatus as would facilitate this arrangement.

The Marconi Company had always steadfastly opposed any such exchange of messages between competing systems; the Allgemeine Electricitäts-Gesellschaft had always furthered the idea. This action of the German delegates to the Conference brought the conflict between these two companies into the realm of international politics; it was said in Britain that Germany was attempting to do by diplomacy what she had failed to do in business competition. The representatives of England and Italy—Marconi strongholds—dissented to the proposal, and though a Convention was nevertheless drawn up, it failed of ratification by sufficient governments to make it of any importance.

Three years later Germany tried again, and again the nations of the world sent delegates to Berlin. By this time it was really necessary to arrive at some sort of agreement—even the controversial "exchange of messages" question could not stand in the way.

An international Convention was signed that agreed, at last, to this old plea of Germany's. Apart from that, it set up an International Bureau at Berne, Switzerland, for the collection and dissemination of the required information regarding the various systems, stations in each country, etc.; it pledged the nations to avoid international interference and to give priority to distress calls; as before, it settled the question of radio rates.

Supplementary to this Convention, Operating Regulations were adopted to cover the telegraphic code to be used, the method of calling and answering calls, etc. With these regulations the distress signal "SOS" came into being. "CQD" had been the Marconi Company's signal for that purpose since February 1st, 1904, and had been the call generally adopted, but "CQD" contained more "dashes" than "dots" and consequently took a longer time to send than did the "SOS," which was proposed by the Germans as a substitute and adopted by the Conference.

The Convention and Regulations were to go into effect on July 1st, 1908. In the interim they were submitted to the various governments for ratification. There was still opposition in England to the "exchange of messages" clause, and the British government and the Marconi Company were sufficiently in accord to prevent that nation from signing by the latter date. The matter was finally compromised when the Government agreed that if the Marconi Company showed a loss as a result of the Convention, the Government would make it good by subsidy for a period of three years; at this the Marconi Company acceded, and the Convention was ratified.

Only the United States and Montenegro, of all the nations participating in the Conference, failed to ratify the Berlin Convention. The United States government, remote from the controversies which had made agreement essential in Europe, and with its customarily lethargic attitude toward international pacts, yielded to the pleas of the commercial wireless interests on this side of the water, all of whom opposed ratification (though the Navy advocated it) and did not sign. The whole matter was before the Senate many times, only to lapse again and again—not until 1912, when a third International Conference was imminent (this time in London) did that body agree to the then six-year-old Convention. It was signed in that year in order that we might be represented and vote at the coming deliberations.

On January 23rd, 1909, the White Star passenger liner *Republic,* proceeding through a thick fog off Nantucket Island, was rammed by the Italian *SS. Florida* and immediately commenced to sink. The *Republic* was fitted with Marconi wireless, and her operator, John R. Binns, broadcast a "CQD" which brought rescue ships and resulted in the saving of the lives of all but six of the endangered passengers and crew.

This was not by any means the first time that the use of wireless had brought about a major sea rescue—during the decade in which the art had grown there had been a dozen such—but this particular episode, occurring so near a metropolis like New York, had elements of suspense and drama that made it eagerly devoured "news"; it was a perfect demonstration of the value of radio at sea and provided the best publicity the art had yet had as a saver of life. The Marconi shore station at Siaconset, R. I., caught Binns' messages and those of the rescue ships, and gave them to the New York newspapers and to the Press Associations; the whole public, tensed to the race against time, dramatized the deck and lifeboat scenes and breathed relief when the rescue was accomplished. Binns, who was among the last to leave the ship, became a national figure; wireless the godsend of the age.

There were in all nineteen marine disasters in which radio figured in that year 1909—but 237 non-equipped ships also foundered, and the loss of life in them seemed futile and unnecessary now that wireless was available. Feeling about the situation found expression in reform legislation. On June 24th, 1910, President Taft signed an Act of Congress which declared it unlawful after July 1st, 1911, for any United States or foreign passenger vessel carrying fifty or more persons, including passengers and crew, and plying between ports more than two hundred miles apart, to leave or attempt to leave any port of the United States unless equipped with "an efficient apparatus for radio-communication, in good working order," and in charge of a person skilled in its use. A fine of five thousand dollars was prescribed for violations; the Secretary of Commerce and Labor was put in charge of carrying it out.

Valuable as this legislation was from a standpoint of safety of life at sea, it was not completely effective because it failed to recognize the old basic principle that it availed nothing whatever to have a sending station if there existed no receiving station to complete the communications circuit. The 1910 law made it possible for any ship to broadcast an appeal for help, but since it required the employment of but a single operator and failed to specify a constant watch at the receiving instruments, it was not entirely adequate.

On April 14th, 1912, the 45,000-ton *Titanic,* largest and most luxurious ocean liner built up to that time, and the White Star Line's entry in the competition for supremacy on the Atlantic, was three days out from Liverpool on her maiden voyage, bound for New York carrying 2223 passengers and crew—among her passengers were listed some of the most distinguished men and women of Britain, Europe and America. On the night of the 14th she ran into the fog peculiar to the vicinity of icebergs at sea, but because of the desire that this first crossing of hers be in every way a record one, did not slacken speed. She was making more than twenty knots when, at ten o'clock, a gigantic berg loomed up in the foggy dark on her starboard bow; unable to change course in time, she struck it and rode up over its huge underwater portion, the impact literally taking the bottom out of her. Watertight subdivision kept her afloat for four hours, but she listed so that fewer than half the lifeboats could be lowered. When, at 2:20 A.M., on the morning of the fifteenth, she sank, more than two-thirds of those who had been aboard went with her, or, jumping as she went, were carried down by suction . . . 1517 perished; 706, huddled in boats or clinging to driftwood, were rescued at dawn.

The distress calls broadcast by *Titanic's* Marconi wireless were heard over a wide area, and of the ships that caught them, *Carpathia* and *California*, the nearest, made full speed for the scene, and *Carpathia*, first to arrive, picked up the survivors. The essence of the tragedy developed later, however, when it was learned that there had been a ship much nearer than either of those two, a ship equipped with wireless but with only one operator and that operator off watch. . . .

The *Titanic* had been considered unsinkable; her loss, with its attendant dramatic stories of much heroism and some cowardice, brought to the public a sense of cataclysm and resulted in searching inquiries which effected many changes in maritime law. The Radio Act of 1910 was amended on July 23rd, 1912, to require two operators and a constant watch, and—since *Titanic's* radio had failed when water reached the ship's generators—an auxiliary source of power capable of operating the wireless for four hours. Voice tube or other efficient means of communication between the radio room and the navigating bridge was required, and the scope of the Act was extended to include cargo vessels.

The passage of the Act of 1910 was of course followed by a great increase in the use of radio, but since there was nothing in the laws of the United States to regulate that use, the efficacy of the newly established communication threatened to become nullified by interference. Any station could use any wave length, could tune "broadly" or "sharply" as its wish might be; the Marconi Company was using International Morse code, some of the American companies were using American Morse; amateurs were increasing in number—in 1912 there were nearly a thousand of them, and their organization, the American Radio Relay League, was five years old.

Under this state of affairs it was no longer possible to procrastinate on the long-considered regulation of radio-communications, either as to joining the international movement or to erecting some sort of regulatory structure of our own, particularly since we had led the world in inaugurating laws requiring the use of radio at sea.

On April 3rd, 1912, then, the Senate consented at last to ratification of the Berlin Convention of 1906. This action entitled the United States to representation at the forthcoming London Conference, which was to open on June 4th. A delegation sailed, and on July 5th signed the new Convention which brought the former agreements up to date. It was ratified by the Senate on January 22nd, 1913.

Even apart from our own muddled conditions, this concurrence in the International Conventions required legislation to enforce the Regulations which were a part of them, and on August 13th, 1912, An Act to Regulate Radio-communication had passed both houses and was approved by President Taft. It provided that before any radio station could operate it must have a license from the Secretary of

Commerce and Labor, and that such stations could be operated only by qualified operators, similarly licensed. Attached to it were a number of Regulations intended to prevent interference, etc., and for the first time an allocation of wave-lengths was made; 300 meters and 600 meters were designated for ship use, amateurs were required to stay below 200 meters, and the band between 600 and 1600 was set aside for the government.

This law made, too, a half-hearted attempt to establish a principle of American ownership of radio stations in this country. It required that station licenses be issued only to citizens of the United States or Porto Rico, but proceeded immediately to abrogate that dictum by adding: "or to a company incorporated under the laws of some state or territory of the United States, or Porto Rico," which provision made it possible for any foreign agency to establish a subsidiary company here and thus acquire the right to operate.

The new law was less a specific radio law than it was an extension of the already existing Navigation Laws to cover the radio field. However, the Congress recognized that there might grow to be more to the art than its seafaring application, and made the law broad enough to cover both transoceanic communication and radiotelephony, so that as these branches of the art developed there was some, though an inadequate, statutory control over their operation.

Tucked away insignificantly in one paragraph was the statement that every station license was to contain a provision that in case of war, the station might be taken over by the Government, and adequate compensation made to its owners. It did not seem, in 1912, as though it would ever be necessary to do anything like that. . . .

Even before the passage of the Act of 1910 requiring ships to be equipped with radio, a majority of the first-class vessels had been so outfitted, but the law was, of course, a boon to the radio companies. They no longer had to "sell" the radio idea. 1912, therefore, saw the maritime branch of the art reach extensive development, saw, indeed, the arrival of Marconi "spark" wireless at its peak. By the end of that year the American Marconi Company's stations ranged at intervals along the full length of the Atlantic seaboard, the Gulf of Mexico, the Great Lakes and the Pacific states. At sea, Marconi wireless was literally the only wireless, except for ships of the United Fruit Company, and through its patents and its sound operation, the company had virtually achieved a monopoly.

This condition had not been arrived at without competition. In the early years there had been several small American companies; in 1907 most of these, including the old American De Forest Wireless Telegraph Company, had been combined to form the United Wireless Company, which had since proved to be an ardent rival.

American Marconi had had, of course, all the advantage in equipping and working with American vessels engaged in foreign trade, for by 1912 the British Marconi Company had made a strong beginning toward the world-wide wireless to which it aspired, and had, besides its American and Canadian companies, subsidiaries in Belgium, France, the Argentine, Russia, and Spain, so that Marconi service offered communications in many lands. Likewise, American Marconi had gotten the American business of most European trans-Atlantic liners.

In the coastwise trade, however, the United Company suffered no such handicap. It built a number of stations on both coasts, and by active salesmanship and cutting Marconi prices—which it was able to do by making free use of Marconi patents without having incurred expenses either for development or royalties—worked up an extensive traffic with coasting steamers.

But the United Wireless Company suffered from a failing peculiar to several of the early American wireless companies—its financial policy was distinctly shady. Apart from its patent infringements, it was engaged in Wall Street activities that had other ends than efficient communication service. In June, 1911, its President and various other officers were convicted of selling stock under false pretenses. Moreover, the Marconi Company had brought suit on the patent question; United pleaded guilty to infringement, and a permanent injunction was issued, restraining them from further use of the basic "spark" and "tuning" principles.

Under these circumstances the company went into bankruptcy with a crash. Its investors faced a total loss and its wireless system, which, notwithstanding conditions at its head, had been rendering real service, was threatened with elimination, leaving a large portion of the American merchant marine suddenly and involuntarily deprived of radio.

This was a tragedy to be averted if possible. In March, 1912, then, Mr. Godfrey Isaacs, representing the British Marconi Company, made a trip to the United States to straighten matters out. Through him the British Marconi Company purchased the entire assets of the United Wireless Company, and on the twentieth of that month resold them to the American Marconi Company, which took over all its operations and business. The stock of the many United shareholders was exchanged for Marconi stock, and thereafter the American Marconi Company was virtually alone in the ship service field; from that time until the war it handled 95 per cent of the marine wireless traffic of this country.

There was one large American shipping company, however, which from the beginning had subscribed neither to United Wireless nor to the Marconi Company's service, but had built up and maintained its own radio-communications system. This was the

United Fruit Company, engaged in the business of growing bananas and other tropical products in Central and South America for shipment to the United States and abroad. It operated the "Great White Fleet" of some twenty-five specially built ships to bring its products north. Its Caribbean plantations were for the most part poorly served by cable or not at all, and since the perishable nature of fruit cargoes made it most important that its business offices have advance information of the exact date of its ships' arrivals, the Company had begun in 1904 to experiment with radio.

The first United Fruit shore stations, at Limon, Costa Rica, and Bocas del Toro, Panama, were equipped with 1904 De Forest sets. Between that time and 1908 the company outfitted all its ships and built six more shore stations, in Nicaragua, Guatemala, Cuba, Louisiana and on Swan Island in the Caribbean, buying equipment from various small manufacturers.

Constant improvements were made and much money was spent in attempts to establish satisfactory communications in spite of vile operating conditions and very bad static. In 1912, tired of depending upon others for apparatus that never lived up to expectations, the company decided to manufacture its own, and purchased a controlling interest in the Wireless Specialty Apparatus Company of New York, which seemed, because of its ownership of the crystal detector patents, to be in a better strategic position than most of the other small manufacturing companies.

Crystals of certain minerals which had been found to possess the quality of "rectifying" radio waves, were the best detectors then on the market; the crystal device was far superior to the electrolytic rectifier which had been its immediate predecessor. It had been Mr. G. W. Pickard who had in 1907 discovered and patented the silicon form of this detector, and the Wireless Specialty Apparatus Company was his company, formed to exploit this invention. In the years intervening he had made a number of improvements, and the crystal detectors which he sold were replacing all the earlier forms of detector. But Mr. Pickard was willing to sell out to United Fruit because the day of the radio manufacturer had not yet arrived—the communications companies dominated the field.

With its own manufacturing company, the United Fruit Company radio service was now on a more satisfactory basis, but one other change was still desirable. A subsidiary company was formed with no other purpose than the handling of radio—in 1913 the new Tropical Radio Company took over the operation of all the ship stations and almost all of those ashore.

During these years in which wireless was becoming part and parcel of the maritime scene, the inventors and developers were busy in their laboratories and several important innovations came forth from them. In 1903 a Dane, Valdemar Poulsen, developed an "arc" transmitter better than any "arc" before. . . . Marconi, in 1902, invented a magnetic detector that rectified, but in 1904 Professor Fleming, one of his consulting engineers, made an adaptation of a discovery of Edison's and produced a vacuum tube detector with two elements in it—a hot filament and a cold plate—that did the work better than Marconi's. . . . In 1907 De Forest conceived the brilliant notion of adding a third element—the grid—to this tube; he called his invention an "audion" and the Marconi Company sued him when he tried to use it, claiming that the Fleming patent covered it. They won their case, but the forerunner of the modern vacuum tube had been devised. . . .

De Forest made radiotelephone sets and sold them to the Navy; they went around the world with the Fleet in 1908. . . . Fessenden worked with telephony, too, and American Telegraph and Telephone Company engineers came to his demonstrations. . . . In 1907 the Navy established an experimental laboratory for radio in the Bureau of Standards at Washington. . . . That Bureau later made important contributions to radio-measurements. . . . After the 1910 Radio Law was passed, the Department of Commerce instituted an Inspection Service to see that its provisions were carried out; this became a model service and its organization was copied all over the world. . . .

And attacks on the broad Atlantic continued ceaselessly. The attempt was to break it down by sheer force, to create machines and antennae so huge that the energy flung into the ether would crash down the distance barriers in spite of any opposition nature could offer.

The Poldhu-Cape Cod and Poldhu-Glace Bay circuits having failed of the desired result, in 1905 Marconi started work on a giant station at Clifden, Ireland; it was nearly two years in building. It had a new type of antenna, larger than any before (because of the increased power and longer waves which it was to radiate), and of a different shape. Eight tall masts, set in pairs, supported fifty-two wires each a thousand feet long; the antenna was two hundred feet above the ground and two hundred feet wide. It led straight for Glace Bay, the end at which the wires converged into the operating house being nearest that faraway point, and it had "directional characteristics"—to a certain degree it pointed the signals toward their destination.

The Glace Bay station was improved correspondingly, and in 1907 these two were exchanging messages with better results than hitherto, though they were continually subject to interruption by static and general weather conditions. Indeed, results were good enough to decide the company that the time had come when the stations might be opened to certain forms of service in competition with the cables. Press messages were first accepted, and then on October 17th trans-Atlantic "Marconigram" service was begun for the general public; the next day 1400 words were transmitted from Clifden to Glace Bay.

The opening gun had been fired. For a quarter of a century the cable rates between London and New York had been twenty-five cents a word; now the radio rate was established at eighteen cents. But the attempt did not develop anything that could be called competition. The cable was fast, capable of handling all the traffic, and above all, it was certain—messages came through immediately, without mistakes. Marconi's hopes for his new stations were not fulfilled by their record in operation; on some days they could do everything the cable could, on others they would be helpless for hours on end, their messages delayed interminably.

On a trip to the Argentine in 1910 the Italian received Clifden 4000 miles by day, 6735 by night . . . yet that station could not operate across the Atlantic with cable regularity. Naturally, the cables continued to get the business, regardless of the difference in rates.

Fessenden, too, assailed the Western Ocean. In 1905 the National Electric Signalling Company transferred its experimental operations to Brant

Rock, Massachusetts, and erected a big station there, with a 10-kilowatt "spark" transmitter, an umbrella-like antenna supported by a 420-foot tubular tower, and the Fessenden electrolytic detector in a tuned receiving circuit. The company had applied to the British Post Office and been granted permission to build a station at Machrihanish Bay, Scotland, for experimental transoceanic communication; work on that station went ahead simultaneously with the Brant Rock installation.

On January 1st, 1906, both stations were ready, and the first messages were exchanged. To Fessenden's delight, reception was excellent; for three days he thought success was his as the two stations exchanged message after message, but then unaccountably communication failed. Try as he might with the same instruments that had lately been so successful, not a sound got through. It was three weeks before communications were established again—under such conditions they labored all year. There would be short periods of triumph, longer intervals of despair, until toward the year's end a winter gale swept down upon Scotland and wrecked the Machrihanish station, terminating the venture.

The indefatigable Fessenden, dissatisfied with the "spark" apparatus, wanted to develop his alternator before building again. He had great faith in the "continuous" waves, and in the alternator as a means of their generation. Even while his "spark" apparatus was working with Machrihanish, he had had his men working on the construction of the first of the new machines which would, he hoped, replace the older devices with happier results. In his shop they wound the field coils and armature and built the electrical end of it, then sent it to the DeLaval Turbine works to have a steam turbine added that would send it spinning at the necessary dizzy speeds.

But speed makes heat; heat destroys bearings, and bearings, both in a turbine and in an alternator, must be perfectly aligned or the machine will tear itself to pieces. Fessenden's first alternator failed. He tried another, but this time instead of building it himself, sent his plans to the General Electric Company, at Schenectady, New York, and in September, 1906, that company delivered a 2-kilowatt machine built to his specifications. This, too, failed. Fessenden tore it apart, had his men rewind and readjust it, and after several months got it working, but delivering less than half the power it was designed for.

Between then and 1911, constantly changing and improving his designs, he got about twenty alternators from the General Electric Company, all small, experimental, and none of them really satisfactory, though the building of them contributed a great deal to the general knowledge of such devices and the problem of their use and manufacture.

Fessenden, however, did not limit his work to alternators. He was interested in all forms of Hertz wave generation, and in 1909 made a valuable improvement to the "spark" apparatus when he developed the 500-cycle rotary synchronous spark-gap. This generated "damped" waves, but waves less damped than those of an ordinary open spark-gap. They oscillated at a high frequency, had a pure note, and penetrated static well. Above all, the rotary feature kept the gap from heating and made it possible to use higher powers, thereby getting better distance.

The alternator was not commercially practicable, but this new spark-gap was. A number of powerful transmitters equipped with it were sold, some to United Fruit, some to the Navy, for both ship and shore use; one of the latter was a 100-kilowatt transmitter made in 1909 for the proposed Navy Arlington station. The National Electric Signalling Company began to be a factor in the radio manufacturing field.

But the company wanted to do more than manufacture. It wanted to get into the trans-Atlantic communications game. It proposed, now, to establish both England-Canada and England-United States circuits, and in 1910 applied again to the British Post Office for licenses. Professor Fessenden himself went over to conduct negotiations.

The Post Office required, as proof of the company's ability to communicate across the sea, that they carry out satisfactory tests between Brant Rock and the newly Fessenden-equipped United Fruit Company station at New Orleans, 1800 miles away, holding this distance across land to be equivalent to the greater sea distance. If these tests were satisfactory, they offered a nine-year license.

Messrs. Given and Walker, backers of the company, thought this term too short to warrant the investment that trans-Atlantic stations would require. They wrote back to Fessenden, still in England, who took the matter up again, and in December succeeded in getting an extension to fifteen years. He sailed for home, carried out the Brant Rock-New Orleans tests . . . and then the whole project crashed into the rocks of internal dissension when the engineer and his backers split wide apart.

The point at issue was a proposed Canadian subsidiary, which was to operate the Canadian end of the Canada-England circuit. Professor Fessenden, born in Canada, had felt that for reasons of propriety—this circuit being an all-British affair—the Canadian Company should be controlled by himself and his British and Canadian associates to the exclusion of Messrs. Given and Walker—who were, however, to advance part of its cost as a non-voting investment—and had secured an agreement from those gentlemen to that effect before he started for England to conduct the negotiations.

This may have been patriotic, and a splendid idea from the British standpoint, but as time went on it seemed that between the Professor and his British associates, the whole venture was becoming one in which the Pittsburgh bankers, who had sunk nearly two million dollars in financing the Fessenden experiments and who controlled the National Electric Signalling Company, were in danger of emerging on the short end of things.

Bad feeling and suspicion between them and their engineer became so great that it was impossible for them to continue together. Fessenden maintained that they were trying to "freeze him out"; he made a British-American issue of it, quit the company in high dudgeon, and brought suit for damages. In July, 1912, a jury rendered a verdict in his favor and allowed him $406,000—the straw that broke the camel's back. . . .

The company, wishing to appeal, could not do so without either filing a bond for that amount or going into bankruptcy. It chose the latter course and from that time until the outbreak of the war operated in the hands of a receiver. The appeal was taken and the case dragged on through the courts . . . ultimately a compromise cash settlement was made, but that was long afterwards.

So ended America's first serious attempt at trans-Atlantic radio communication service, leaving the Marconi Company virtually alone in that field as it was in the farther-advanced ship service. The end of July, 1912, saw the Marconi Wireless Telegraph Company of America solidly and firmly entrenched as the single consequential operator in the commercial radio field in the United States. It had never paid a dividend, but by virtue of its communications ideal, its Marconi apparatus and its alliance with the world-wide Marconi system, it had held its head above water through the pioneer period that had wrecked so many other radio concerns.

PART TWO

THE ERA OF MILITARY USE

CHAPTER FIVE

HIGH POWER CHAINS

THE second decade of the twentieth century was a military decade. The first had seen a huge growth in international commerce and international competition as the machine gave birth to mass production. But competition is one form of war . . . and commerce had not stopped at business war. During that first decade the manufacturing nations of the world had been arming themselves with battleships, with guns; men by the million had been trained until there was everywhere a vague sense of impending combat. Even America, remote from the European hotbed, was touched by the spirit of the times and joined the parade of sea-arms; our Navy grew with England's and Germany's.

These military trends had a powerful influence upon the growth and development of radio.

In 1909 the British government took over all the ship-service shore stations of the British Marconi Company and of Lloyds. This placed the British shore end of ship-shore telegraphy under government operation, as land-wire telegraphy was already. The stations on ships continued to be operated by the Marconi International Marine.

The great network of cables that ran from England all over the world was chiefly under the private ownership of three associated companies called the "Electra House Group"—The Eastern Telegraph Company, Ltd., which operated to Europe, India and the Far East; the Eastern Extension Australasia and China Telegraph Company, Ltd., which operated to the Australian archipelago, East Africa and China; and the Western Telegraph Company, Ltd., which operated to the Americas and West Africa. However, the State had one cable circuit of its own; this ran under the Atlantic to Canada, connected there with the land-wire telegraphs to the Pacific, where its second branch crossed that ocean to Australia; it was managed by an inter-dominion body known as the Pacific Cable Board.

Commercial overseas radio, when it developed to the point of practicability, was presumably to be —like the cables—in the hands of private companies, but when, in March, 1910, the British Marconi Company applied to the Colonial Office for exclusive licenses to build a chain of eighteen high-power stations linking by radio the British dominions on five continents, the plan met with opposition. The Government refused to grant a monopoly of such vast scope to a private company, and announced its belief that a project of such magnitude should be in the hands of the State if it were entered into at all. The licenses were denied, and upon a second application a year later this attitude was reaffirmed.

Then, in April, 1911, the Marconi Company submitted another plan. By this time they had developed "spark" radio—the radio of "damped" waves—to a state of considerable perfection; their position was that since they felt it to be both practicable and desirable to link the British dominions by wireless, and since the State had signified that such an idea should only be carried out under Government auspices, they proposed that it be now entered into, and that they be given the contract for the construction of the stations.

Such matters in the British Empire required the coöperation of the Dominions; the Government called an Imperial Wireless Telegraph Conference to consider the whole subject and make recommendations.

They met in London, the representatives of all those lands upon which the sun never set and which were already so elaborately interconnected by that cable system the erection of which had been one of the marvels of the nineteenth century . . . and they decided that "the great importance of wireless telegraphy for social, commercial and defensive purposes renders it desirable that a chain of British State-owned wireless stations should be established within the Empire."

Why was this? Radio had not yet shown itself ready to replace the older transoceanic telegraphy, or even to stand beside it.

The reason was military. It was a time when defenses were being tightened. Inter-dominion communication in every form—by ships, by post, by telegraphy—was the life-blood of the British Empire. In case of war the enemy might *cut* the cables. Therein lay the vulnerability of the existing telegraph system, and the thought of being without telegraphic communication, particularly in time of war, was enough to make one shudder. By all means, build the radio stations.

Thus, as safety of life at sea had forced the adoption of radio telegraphy for "insurance" purposes even while its maintenance was a dead loss as a revenue producer, so, in the second era of the infant art, it trended toward adoption for transoceanic communication for "insurance" some time before it was ready to stand on its own feet as a commercially paying proposition when so used. Both circumstances redounded to the benefit of the radio companies and the advancement of the art, for, like the shipowners, governments were now willing to spend millions of dollars to build and maintain stations merely that they might be ready;

they demanded, likewise, constantly improved equipment.

Britain accepted the Conference's recommendations, and early in 1912 signed a contract with the Marconi Company for the erection of the first six stations—in England, Egypt, East and South Africa, India and Singapore—of "The All-Red Chain," as it came to be called. Marconi stock skyrocketed until a one pound share was selling for nine—the company was to build and operate the stations for the Government; each station was to cost £60,000; the company was to receive 10 per cent of the gross receipts as royalty; the term of the contract was twenty-eight years, though it might be ended by the Government after eighteen under certain conditions. . . .

But things were not to be so simple.

Here was a contract involving millions of pounds-sterling, the benefits of which, though plain to the initiate, were intangible to the layman. There are always those who object to "insurance," and in this case they up with the loud cry that the plan was silly and wasteful. It was said in Parliament and in the press that the government men who had granted such a plum to the Marconi Company had done so for corrupt and ulterior motives, that the real secret of their action was to be found in the stock exchange—that they had bought Marconi stock when it was low and were now selling at the top of the high market caused by their own actions.

Work was halted, and in January, 1913, Parliament appointed a committee to reconsider the plan and decide upon its wisdom and on the probity of the involved officials. The committee met; it witnessed, among other things, tests between the Marconi stations at Clifden, Ireland, and Glace Bay, Nova Scotia; it heard both sides of the question, and it was convinced that the idea of a State-owned All-Red Chain was sound and ought to be adopted. Furthermore, it recommended the Marconi system as the best for the purpose and exonerated the gentlemen who had made the contract.

A new contract was drawn up by the Post Office in July, and ratified by the British and Dominion Governments; its features were substantially the same as those of the first.

While all this was being done there was a fresh cycle of Marconi wireless attacks upon the Atlantic, and this time, upon the Pacific as well.

The most determined effort yet was being made by the Marconi men. With the increase in power gained by using the latest "spark" developments —the "quenched" and "rotary" gaps, improved antenna design, larger masts and aerials and longer wave lengths—they were building again. The circuit was to comprise four stations, one on each coast for sending, and one on each for receiving. Carnarvon, Wales, was to send to Belmar, New Jersey; New Brunswick, New Jersey, was to send to Towyn,

Wales. This would make possible the transmission of messages as soon as they were filed, whereas formerly, since stations could not send and receive simultaneously, they had had to wait until the air was clear.

At Marion and Chatham, Massachusetts, another such pair was begun for direct communication with Stavanger, Norway.

But in spite of dreams and ambitions, none of these stations got into service for the purpose intended; like so many other plans, these were interrupted by the declaration of war in Europe. In August of 1914, as the preliminary tests were being carried out between New Brunswick and Carnarvon, England commenced her great effort and the British Admiralty took over the Carnarvon and all other high-power Marconi stations.

The All-Red Chain, too, went by the board with the coming of war, and the activities of the British Marconi Company, which had been expanding so rapidly, were virtually ended until peace should return. The masts for the stations in England and Egypt were already up, and some of the material for that in India had been delivered. Long afterwards, in 1919, the Marconi Company was awarded damages for this breach, and £500,000 were paid as "royalties that would have been earned if the agreement had not been terminated." . . . In the meantime, the company's men, stations and resources went into service, and British radio joined the colors.

Only the Marconi International Marine was active; because of the huge war shipping, that company had everything on its hands that it could handle.

The American Marconi Company, halted for the time in its trans-Atlantic activities, nevertheless had the Pacific for a field. Simultaneously with the start of its big East Coast stations, it had begun stations in San Francisco and Honolulu; on the 24th of September, 1914, this United States-Hawaii circuit was opened to the public in competition with the cable of the Pacific Commercial Cable Company; the rate was fixed at twenty-five cents a word, as against the cable company's thirty-seven, and from the beginning the new service was popular—the cable company met the cut in rates.

On July 27th, 1915, a contract for the exchange of messages having been signed with the Japanese government, this Marconi Pacific service was extended to the Japanese station at Funabashi, near Tokio; radio messages could be sent, then, from the United States to the Flowery Kingdom for eighty cents a word, forty-one cents less than the former cable toll; ultimately the two rates stabilized at seventy-two cents a word for radio, ninety-six cents a word for the cable.

The Pacific was an attractive field. For one thing, it was not nearly so well served by the cable as was the Atlantic; for another, radio operating conditions were better there than they were on the older ocean.

There was less "static," less of the mysterious "fading" that was so maddening on the other coast. Last of all, and most important, the Pacific had a huge trading future; it was virgin territory; it was a place in which a company could grow with the times.

Even before the Marconi Company commenced Pacific operations, another group of men was surveying the field. As early as 1910 they saw possibilities not only of communications to the Orient, but also for radiotelegraphy between the five metropolitan cities of the West Coast—large cities, but separated by great distances, so that the maintenance of wire lines was expensive and rates correspondingly high. As far as the ship-shore business was concerned, that field was limited, for it was already well served by United Wireless (later taken over by the Marconi Company). The best field, then, from a revenue standpoint, would be in high-power chains, and with such ideas in mind, they cast about in that year to see what was available in the form of high-power apparatus.

There was not much. That of the Marconi Company, a prospective competitor, was of course denied to them, and a more or less similar situation held with the National Electric Signalling Company. In any case, they did not want to be in the position of paying royalties for the use of whatever apparatus they decided upon, so they turned to Europe to see what was available there.

They found the Poulsen "arc converter," invented and patented seven years earlier by the Dane, Valdemar Poulsen. This was a generator of "continuous" waves; it had two distinctive features that marked it from the ordinary "arc,"—in the first place its "arc" burned in an atmosphere containing hydrogen, and secondly, it burned in a strong transverse magnetic field. As used, it was included in a circuit containing inductance and capacity which it caused to oscillate at radio-frequencies; it was capable of development to very large sizes.

In October of that same year 1910, these men formed the Poulsen Wireless Corporation, incorporated in the state of Arizona for twenty-five million dollars (an exaggerated capitalization, later cut down), to buy the United States rights to the Poulsen patents. This they did, for the sum of half a million dollars.

To commence operations, they would have to manufacture apparatus, erect stations, and set up a communications service. In February, 1911, they formed the Federal Telegraph Company of California for these purposes, and commenced the struggle that had always been a feature of radio development—that of bringing a fundamental idea to its full utility in practical operation. There was this about their "arc," however,—it had a good deal of power to start with (their first apparatus was thirty kilowatt), and its continuous wave principle meant that more of this power would get to the antenna and into the ether than in any "spark" development of similar rating.

They hired engineers—among them Lee De Forest, just severing his connection with United Wireless—built experimental stations at El Paso and Fort Worth, Texas, and at Chicago, and went ahead with activities. In 1912 they were ready; that year work started on stations in Los Angeles, San Francisco and Portland. By 1914 they had erected a Honolulu-San Francisco circuit in competition with American Marconi and the cable, and from that time on they were a decided factor in the communications of Western America.

This was the first substantial American radio company.

The "arc" was destined, however, to play a larger part in American radio than in its commercial communications on the Pacific.

1912 had been a large year in radio. The Congress in that year was "radio-minded"—it amended the 1910 radio law, passed the Radio Law of 1912, ratified the 1906 Berlin Convention, and in general behaved in an altogether constructive manner. That was the year of the All-Red Chain; the United States Navy, led in similar direction by the old Report of the Roosevelt Board of 1904, had been trying for several years to establish a similar high-power chain linking this country with our Pacific island possessions, and in 1912 Congress finally authorized them to go ahead to the extent of a million dollars. The United States, as well as Britain, was "insuring" in those years. It was to have an "All-Red Chain" of its own.

Navy radio had expanded considerably since the pioneer days. The first few Navy stations along the North Atlantic coast had been added to in the decade intervening until they dotted both seaboards and their masts stuck up at isolated places in Alaska, the Caribbean and the Pacific . . . in Porto Rico and the Canal Zone, the Privilof Islands . . . in Cavite . . . and every Navy ship carried wireless.

Like many another similar system, this one had "just growed." For several years after its first installations, it had stuck to Slaby-Arco apparatus, buying improved sets from Germany as they came out, and making still other improvements by tinkering to adapt the crude instruments to service conditions. Then, with the growth of the National Electric Signalling Company and of De Forest's manufacturing company, it had turned to those organizations for its needs, and to such smaller companies as Pickard's Wireless Specialty Apparatus Company.

All this was really experimental work—education by the trial and error method. It was carried on by the Bureau of Steam Engineering—formerly the Bureau of Equipment—which handled all electrical apparatus, and because of the circumstances, Navy radio apparatus covered about as wide a range of utility, from good to bad, as there was in the world.

It was used chiefly for communication between ships and shore; to a lesser extent between ships out of sight of each other—ships in squadron still exchanged their messages by signal flags and semaphore.

The Navy faced a growing problem in radio operation. So far, ships had simply been given sets and left to use them very much as they saw fit; for operators they had had naval electricians and signal-quartermasters adapted to the new work, . . . and though communications was a field of its own, an art, what those operators knew about it was in the main self-taught and often full of those novel ideas peculiar to the bluejacket the world over.

The whole situation was somewhat similar to that in which the United Fruit Company had found itself; the solution was identical. A separate operating branch was formed, which in the Navy was called the Naval Radio Service; it was given a Superintendent, and that officer was delegated to go ahead and organize it, draw up regulations, operating procedure, etc., to make the new service efficient.

This was infinitely better than the old more or less unguided state of affairs, but it still left undetermined certain matters of policy that would, as events began to demonstrate, soon have to be settled. The 1912 law, for example, indicated that in case of war all private radio stations were to be taken over by the government (which meant by the Navy), and in 1913 and 1914 things were happening that were turning the minds of military men toward such possibilities. What would be the plan of procedure for radio in such case? How would messages be gotten from Washington to ships? Which stations would be used for what?

The Superintendent of the Radio Service submitted his recommendations in the matter, and after discussion between the Commanders-in-Chief of the Fleets, the Secretary of the Navy, and the General Board, a board was appointed on December 5th, 1914, to reorganize the Naval Radio Service. It reported in February, 1915; its report was approved in May of that year, and Navy radio took new and final form under the name of the Naval Communications Service, this time an adequate and substantial conception. The first Director of Naval Communications was a captain named William H. G. Bullard who had taken an active part in naval electrical and radio developments for many years, and whose two-volume electrical text-book was in use at the United States Naval Academy.

Meanwhile the *materiel* end of the service was likewise being pulled together. Since 1909 the Navy had operated a laboratory in the Bureau of Standards for the purpose of developing radio equipment that would fill the somewhat specialized needs of the service; this had been in charge of Dr. L. W. Austin. There had naturally been, too, a number of officers and bluejackets who had come to know considerable about the art, and now the pooled knowledge of all these men began to make itself felt. Instead of being entirely dependent upon the manufacturers and what was available commercially, the Navy began to have ideas of its own and to submit plans and specifications of what it wanted. There grew up in this way a close degree of coöperation between the service and those who manufactured apparatus for it.

When, in 1912, the Federal Telegraph Company had entered the field, it had come at a time when the Navy was in particular need of high-power apparatus. The giant station at Arlington—a project that had been under way for three years—was nearing completion; its three huge self-supporting steel masts, one 600, the others 450 feet high, were up, and the antenna being stretched. It was being equipped with a 100-kilowatt Fessenden high-frequency "spark" transmitter that had been in storage since 1910 and that was not altogether worthy of a station planned to be "the best in the world."

Before the station's completion, one of the navy men working on it had stretched a single wire from the top of the tallest mast to the ground, and using this for an antenna, had connected up a receiving set. One night in 1912, while "listening-in," he heard a strange meaningless assortment of signals, very faint, come through the air; his hand wrote them down automatically. Presently they broke into International Morse code in a foreign language. When, next night, he listened again, he heard the same signals. An officer told him they were in French.

Upon investigation these messages were found to be time signals from the Eiffel Tower radio station in Paris! But would Paris be able to hear Arlington with the equipment being installed?

There was also the question of the high-power chain recently authorized by Congress; instruments for that would have to be found. . . .

In February, 1913, Arlington went into commission. Very shortly afterwards the Federal Telegraph Company installed a 30 kilowatt "arc" transmitter there for comparison with the Fessenden "spark" apparatus already installed. The two were tested to Panama, which was to be the location of the first station on the chain, and the "arc," though it had less than half the power of the Fessenden machine, "got through" better.

Thereafter the "arc" became the Navy's darling, and that service adopted it for every use possible. One of 100 kilowatts was ordered immediately for the Canal Zone station at Darien; it was delivered, and the station commissioned in May, 1915. The San Diego and Pearl Harbor, Hawaii, stations of the chain were opened a year later, equipped with 200 and 350 kilowatt "arcs" respectively, and shortly thereafter the largest one yet, of 500 kilowatts, was installed at Manila. With the commissioning of supplementary stations in Porto Rico and Guam the chain was complete.

Ultimately the Navy put "arc" transmitters aboard most of the battleships in the Fleet, and in all the more important of the Navy coastal stations. For its time the "arc" was a very useful apparatus; it had some drawbacks, among them the enormous amount of heat thrown off by the flame—necessitating water cooling—and the fact that it emitted a somewhat broad wave and a number of parisitic wavelets (harmonic waves) that blanketed the area around it. But when carefully handled it was quite reliable, and during those years the Navy made extensive use of it and became thoroughly committed to "continuous" waves.

In 1917, at the time of our entry into the war, the Navy owned and operated thirty-five coastal stations, twenty-three more on light-vessels, and had completed the high-power chain across the Pacific; all its men-o'-war were well equipped with radio apparatus and sixteen of them had newly developed radio "direction finders"; attempts to develop radio for aircraft use were beginning to be successful, and there were on hand fifty sets of aircraft radio apparatus capable of working for distances up to one hundred miles. All told, the United States government had spent about twenty million dollars for radio "insurance" between 1899 and the latter date.

There was in those days one other high-power chain in the world. By 1911, German commercial radio activities had become concentrated in the hands of the Telefunken Company, which was both a manufacturing and an operating organization and which entered the lists to provide radio service for ships of the German mercantile marine. For that purpose it needed stations on the Atlantic coast of the United States, and in that year it formed an American subsidiary—the Atlantic Communications Company.

The first activities of this company were confined to the inspection of ship sets, etc., but in 1912 it erected a 60-kilowatt "spark" station at Sayville, Long Island, which operated with ships and very occasionally with the station at Nauen, Germany. The Telefunken Company had been developing an Alternator along somewhat different lines than those on which Fessenden had worked, and in 1913 brought this machine to a point where it was worth installing in high-power stations. Accordingly, late that year work was begun remodeling the Sayville station for regular transoceanic service.

At the same time the Atlantic Communications Company began the construction of a second high-power station on the Atlantic seaboard, at Tuckerton, New Jersey, ostensibly doing the work for a French company. When war was declared in Europe, both these activities became frankly German in character. Work on them was rushed, for Britain had cut the German cables at the very outbreak of hostilities, and Germany was literally cut off from telegraphic communication with the outside world.

They were opened in the fall of that year, and though they did not give service of cable quality, their existence was of incalculable value to Germany; they were doing a rush business immediately. At the same time their existence presented a distinct problem to the United States government because they constantly violated the provisions of our neutrality. For example, the Germans were able to maintain contact with their system of espionage in England by sending radio messages via Sayville to agents in the United States, which were then forwarded to their British destinations by cable. . . . Ostensibly purely a system of commercial communications, the Telefunken trans-Atlantic circuits soon became strongly tinged with the military.

In 1915, after repeated difficulties, a censorship was established in both stations, the task being assigned to the Navy. Even this did not prove satisfactory, and a year later the stations were taken over by the United States government and operated by the Navy. The apparatus at Tuckerton was at the time broken down, and the German Alternator at Sayville survived only two weeks of naval use—possibly the unfamiliar hands of its new operators having something to do with it. The Navy installed "arcs" at both stations. . . .

When the United States declared war the Alien Property Custodian took over both properties; the Navy later purchased the Sayville station from that agent, while Tuckerton passed into the hands of the

Marconi Company. During the period covered by this somewhat unique entry of the Government into the communications business, over $400,000 in tolls were taken in by the Navy and turned over to the Treasury.

CHAPTER SIX

SCIENCE

If the years 1900-1914 had seen a hitherto unparalleled growth in the world's armaments, they had likewise seen developments in industry that were revolutionizing all life upon the earth. Since the beginning of time the cumulative knowledge of the world had added little to man's personal ability to circumvent the forces of nature. For locomotion he had had his feet and his animals; for warmth, and for light through the dark hours of the twenty-four, he had had fire in some shape or form; for clothing he had had skins or the product of the hand-weavers. But in the middle decades of the nineteenth century the growing use of the machine had begun to overcome these conditions and give him greater freedom. Railroads, gas lighting, farm implements, sewing machines, telegraphs . . . the list is epic. Now another revolution was taking place. Man the individual was at last being given personal mastery of nature. He was being given the automobile; the telephone was in every home; electricity was serving him in a thousand ways. Space, distance, discomfort, all were succumbing. The development on the earth's surface was nearly complete; that in the air and "ether" above it was on the verge of completion—the airplane and the radio were well launched. . . .

In the accomplishment of all this the industrial organization had grown to enormous stature, and, furthermore, had been compelled to undergo as sweeping a reorganization and readjustment as had the individual. In no way was this better illustrated than in its relation to science.

Science was the keynote of the era. Scientists led the way with their discoveries of new facts, principles, elements, laws, theories and proofs of theories; "inventors" and engineers adapted or "developed" these basic discoveries to the point of commercial use; industry manufactured the new devices; man, the user, paid for it all and reaped the ultimate benefit.

Of these four steps, the first had long been isolated and remote, dissociate from the rest. Science and its researches had been sacrosant; the scholars and savants who delved into nature's secrets and mysteries had been supported, not by industry, but by education. The Laboratory was a part of the

University, and had been since the beginning of the Renaissance—those early scientific explorers into the radio field, for example, were Professors, teachers; their attitude toward Marconi and the commercializing of radiotelegraphy was typical of the breach between pure science and practical development.

Inevitably, however, this gap tended to close; science and commerce became wedded in spirit as they were in actuality. To begin with, the "scientific method"—that painstaking, dispassionate, thorough investigation which had come to be the scientist's manner of conducting his researches into new fields—was adopted by the engineer-inventor as the best way of carrying on his development work; the Development Laboratory became an important factor in the industrial scheme. And beyond that, as the pace of progress quickened, it became impossible for industry to wait upon the chance discoveries of university laboratories; in the eighties and nineties the chemical and dye manufacturers of Germany and England were engaging eminent scientists and setting up Research Laboratories of their own, whose discoveries were applied directly to commerce in the form of new chemical compounds and coloring matters. In 1914 Germany led the world in these applications.

In 1876 Edison established his development and experimental laboratory at Menlo Park, New Jersey, and in it carried on "invention" by the "trial and error" method, later to use the results industrially; the thousands of experiments he and his assistants made brought to the world, among other things, the incandescent lamp, the phonograph and the motion picture. . . .

His inventions, and those of such others as Nicola Tesla and Elihu Thomson, stimulated an unparalleled growth of electrical application in the United States. The spread of incandescent lighting created demands for generating machinery; the growth of electric street railways and of other uses of electric power was both the cause and the result of constant improvements made in dynamos, motors, insulation, switchboards. . . . Great electrical companies came into being, and inevitably turned to science to keep abreast the rush of the times; the development laboratory became an integral part of the manufactory.

In 1900 the General Electric Company, of Schenectady, New York—then growing to be the largest electrical manufacturer in the United States and an enterprise of world influence and scope—took the next step, and established a Research Laboratory. The particular circumstance which caused this departure was the company's desire to develop a better type of incandescent lamp than the Edison carbon-filament variety which it was at the time manufacturing; it was believed that if pure research in electrochemistry and physics were carried on, an entirely new filament material might be discovered. Impatient business, unwilling to wait, jogged science's elbow.

The Research Laboratory was placed under the direction of a distinguished scientist, Dr. W. R. Whitney, and a group of profound and capable men was assembled to carry on its work; it came to embody the life and soul, the progressive spirit of the company—its workers were literally given *carte blanche,* as time went on, to pursue whatever scientific mystery most attracted their curiosity, for it was the unknown, the unexplored, that beckoned. Within the same building was housed a large staff of engineers engaged in the practical development of new appliances—"inventions"—to be manufactured by the company; these two branches of creative scientific effort worked hand in hand, the engineers forming a liaison between the scientists and the factory.

The venture proved profitable. As time went on the Research Laboratory idea was widely accepted in this country, not only for that reason, but because, as Dr. Irving Langmuir of the General Electric Research Laboratory has said, "of a genuine sense on the part of industry that it was indebted to the pure science of the past and because the modern conceptions of service and the growing *esprit de corps* of American industry helped make it glad of any opportunity to contribute to scientific knowledge."

An idea of what such laboratories as these came to mean to the great cross-section of life in America, and what an intimate relation they bore to every person in the land, may be had from the story of that same incandescent lamp. Between 1900, when the General Electric Laboratory was started, and 1904, the workers at Schenectady made inventions which resulted in the "metallized carbon filament." The lamp containing this improvement was marketed as the "Gem," and consumed only four-fifths as much electrical energy as the old carbon lamp giving the same amount of light. The laboratory next turned its attention to the metal-element Tungsten, which had a higher melting point than any other metal, to see if some means could not be devised for using it as a filament; before their work was completed a German concern developed a tungsten lamp—the General Electric Company purchased the processes and added them to the laboratory's growing store of knowledge.

In 1907 a tungsten lamp was put on the market which, though it was costly to manufacture and hard to ship because of the exceeding brittleness of the filament, used only one watt of current per candle-power of light and made any other lamp seem too expensive to use. The problem of overcoming this lamp's great defect, the filament's brittleness, seemed an insuperable one because the same high melting point that made tungsten so valuable as a filament prevented the metal from being worked as other metals were, by casting, forging or drawing. Nevertheless, Dr. W. D. Coolidge of the laboratories understood the task. He spent many, many months at it . . . and in the end succeeded. He made a drawn-tungsten wire that was stronger than steel piano wire and so flexible that it could be tied into a knot or used to sew on buttons!

The tungsten lamp which resulted from this invention revolutionized lighting. By 1921, 99 per cent of the candle-power produced by incandescent lamps in the United States came from tungsten filament lamps; if the same quantity of light had been produced by he old "Gem" lamps, the electricity used for lighting would have cost $1,255,000,000 as compared to the $455,000,000 which was the country's actual lighting bill in that year. Lamps were cheaper, too—automatic machinery had been developed for making them. . . .

Such savings modern industry brings the world!

In 1909 Irving Langmuir, Doctor of Science, who had been teaching chemistry at Stevens Institute, in Hoboken, New Jersey, and who was later to do some revolutionizing things for radio, decided to spend his summer vacation in the General Electric Company's Schenectady Laboratory. He was a man in his middle thirties, educated in the United States and abroad, and possessed, both by training and inheritance, of a large bump of scientific curiosity; a thorough, persistent worker, a true scientist, and a man of great capacity. Dr. Whitney, recognizing a kindred spirit, suggested that rather than undertake any particular work at once, he spend several days visiting the various rooms of the laboratory . . . he could decide, after that, what would be most interesting as a problem for the summer.

The laboratory was at the time engaged in the development of the new Coolidge drawn-tungsten wire; lamps made with it were working perfectly with direct current, but a curious kind of brittleness called "off-setting" inevitably appeared when they were run on alternating current. Dr. Langmuir thought this might result from impurities in the wire in the form of gases, and since he had been much attracted by the excellent apparatus available for "exhausting" lamps to make a very high vacuum in them, suggested to Dr. Whitney that he would like to investigate this question of the gases contained in and given off by incandescent metals in vacuums. . . .

Thus started a train of experiments which occupied three years, for when fall came Dr. Langmuir was so fascinated by the free hand he had been given in the laboratory that he was only too glad, when Dr. Whitney proposed it, to forego teaching forever, and turn to the laboratory as offering the finest opportunity for pure scientific research and experimentation, the subjects nearest his heart, that he could hope to find. As compared to it, he had come to dread the classroom.

One of the earliest things Dr. Langmuir learned was that if a filament were placed in a glass bulb, and that bulb exhausted by pumping to as high a vacuum as possible (one on the order of 1/1,000,000 of ordinary atmospheric pressure), the high vacuum was lost when current was passed through the filament. This was because there was gas in the filament, gas in molecules, hiding between the molecules of metal, and gas, too, between the molecules of glass in the bulb; when the lamp was heated this gas was driven off until, though to the layman the interior of that bulb would still be a perfect vacuum, to the scientist it was as full of gas as a balloon.

To overcome this condition, Dr. Langmuir came to use the most extraordinary precautions in pumping his vacuums. To drive the gas from the glass of the bulb, he enclosed the whole in an electric oven while it was being exhausted; to drive the gas from the filament he heated that wire until it was almost at the melting point; to prevent any mercury vapor from finding its way into the bulb from the pump, he made a bend in the "lead-off" tube and dipped it into a "trap" of liquid air which froze that vapor and kept it from going farther; he carried on his exhausts for periods of several hours; and apart from all that, he used what was called the "clean-up" effect,—after the lamp had been "sealed off," he heated the filament by passing a current through it, until a small piece of metal, left inside, had been burned; in burning, it combined with any tiny amounts of gas that might be left and eliminated them. The result was to all intents and purposes a perfect vacuum which would remain constant during the entire life of the lamp.

He experimented, too, to see what would happen if he first pumped out the lamp and then deliberately admitted various pure gases to determine which of them had deleterious effects. He found that no two gases acted alike. He learned that it was water vapor that caused the black deposit that formed inside the bulb after a lamp had burned for some time. He found that in a lamp filled with nitrogen at atmospheric pressure the filament could be maintained at very high temperatures (2800 degrees Centigrade) for a far longer time without destruction than if similarly heated in vacuum. This, at last, led to a commercial application—the tungsten filament Nitrogen lamp, with an efficiency thirty or forty per cent better than the vacuum tungsten lamp which had been so revolutionary. And this radical discovery, made possible by three years of purely scientific work, more than repaid his employers for the freedom they had given him and the large sums of money they had allowed him to spend in experimentation. Incidentally, the world had been enriched by much new knowledge about vacuum phenomena. Thus science, industry and business were wedded, to the public's benefit. . . .

In 1911 Dr. Langmuir became interested in what were known as "thermionic" electric currents, inside electric bulbs at low gas pressures (one millionth of atmospheric). It had been known for nearly two hundred years that air in the neighborhood of incandescent metals was a conductor of electricity; in the eighties this knowledge had been added to by the discovery that metals confined in such vacuums as men were then able to pump likewise gave off electricity when heated, and that under such conditions it was always "negative" electricity.

Edison, working with his carbon filament lamp, had noticed such a condition. When he heated his filament beyond a certain point, he could see a "blue glow" between its legs, which suggested the emission of electricity. To study this phenomenon he had enclosed a second electrode in the bulb, and found that the filament *did* give off electricity when heated, that it was "negative" electricity, and that by making the second electrode "positive," the negative filament emissions could be attracted to it. If, however, the electrode were made negative, no current left the filament. This phenomenon, not particularly understood at the time, became identified as the "Edison Effect." It was studied in detail in the nineties, by that same Dr. Fleming, in England, who knew Marconi so well.

Meanwhile a very distinguished English physicist, Sir Joseph John Thomson, was carrying on researches that were revolutionizing man's conception of matter. Hitherto the *Atom* had been the basic and smallest unit known in the physical structure of the universe; Thomson now demonstrated that the atom was made up of other and smaller units which he called *Corpuscles*—they have since become designated *Electrons*. The electron was an infinitely small piece of negative electricity, really an "atom of electricity." Its mass was about 1/1800 of the mass of an atom of Hydrogen. The atom of any substance, then, came to be thought of as consisting of a central "positive" neucleus surrounded by a number of these "negative" electrons, the whole forming a sort of infinitesimally minute solar system.

Thomson thought of the flow of electricity through a conductor as a movement of these electrons between the atoms and molecules of the wire—a very definite physical movement of a material substance. He thought of the electrical discharge in the Edison Effect as the "boiling out" of electrons from the hot incandescent filament; being negative, they were attracted to the other, positively charged, electrode, thus causing the actual flow of electric current across the space between the two.

This view was also held by another distinguished scientist, Professor O. W. Richardson. Richardson further came to believe, with the Germans Reicke and Drude, that in all metals there existed a certain number of these electrons in a "free" condition, able to move about and circulate, just as molecules of gas circulate in space. He believed that these free electrons were ordinarily held within the metal's surface by an electric force, just as the molecules of a liquid are prevented from escaping by a surface force related to the surface tension. If their velocity were increased, as for instance by heating the metal, they would burst through the surface into the space around it.

Richardson believed this action would follow the same laws as those governing the evaporation of liquids; that the number of electrons escaping would increase with the temperature, just as the vapor given off by a liquid increases as its temperature is raised. He found it very difficult to prove this, but gathered enough data to enable him to lay down, in 1903, a theoretical formula for electron emission from heated filaments. This became known as Richardson's Law, and the name he gave such electric currents was *thermionic* currents (from the Greek *therme,* heat, and *ion,* ppr. of *ienai,* go).

Richardson's Law, however, was not generally accepted by his contemporaries, most of whom believed he was wrong when he said the Edison Effect was entirely due to electrons given off from the hot filament. They believed that the currents between the filament and the plate had more to do with the gas remaining in the lamp than with the filament emissions; that this gas became electrified by the filament and carried the current from one electrode to the other. They believed that if all this gas were removed from the tube, leaving the filament and plate in a perfect vacuum, no electrons would be given off. As time went on and other men experimented, this belief became general.

In 1902 and 1903, when radio was first beginning to reach out for great distances, there was a wide search for a better detector than the coherer. What was needed was one that would "rectify" the incoming oscillating wave, so that it could register a direct effect upon a telephone receiver. Fessenden and De Forest, engaged on this quest, developed "electrolytic" detectors; Marconi brought out a "magnetic" type, and Dr. Fleming, who had so long been familiar with the Edison Effect, utilized its principle in the "Fleming Valve," first of the vacuum tube detectors. This was a small glass bulb containing a filament and a metal plate, and exhausted to a good, but not a complete vacuum. When the filament was heated, currents passed across to the plate if that member was positively charged; if the plate was negative like the filament, no current crossed the space between them. It was manifest, then, that if an incoming radio wave were led to the plate, current would flow within the tube only during that time when the Hertz wave was positive. During the other half of its oscillating cycles it would charge the plate negatively and no current would leave the filament. This made the Valve a perfect rectifier, and therefore a detector. It was patented in 1905, and the patent, which covered broadly the use of vacuum tubes as detectors of Hertz waves, became the property of the Marconi Company.

Meantime, Dr. De Forest, dissatisfied with the electrolytic detector and very much interested in telephony, which above all things required a good detector, was looking for something better than had yet been found. He experimented with gas flames; he heard of the Fleming Valves and made some of them with the aid of a glass-blower; he tried everything he could think of or lay his hands on . . . and in 1906 he had a brilliant idea. He conceived the notion of adding a third element to the Fleming Valve, a tiny electrode between the filament and the plate for the purpose of controlling the electron current between those two.

He made such a tube, with the third element in the shape of a tiny gridiron (from which it derived its popular name—the "grid"), and called his device the *Audion*. The third element was all important. Connected to a radio antenna, it became alternately positive and negative as the incoming Hertz waves oscillated; when it was positive, it checked the flow of electron current between the filament and the plate; when it was negative, it stimulated that flow. The Audion was much more sensitive as a detector than was the Fleming two-element Valve.

De Forest used his Audion as a detector both for radiotelegraphy and for telephony.. It was incorporated for that purpose in a number of sets of radiotelephone apparatus that he sold to the Navy in 1907 and which went around the world in 1908 and 1909 when a number of American battleships made that long cruise. In the winter of 1909 he used his Audion to pick up a radiotelephone broadcast which he had arranged from the Metropolitan Opera House in New York, and sitting with a group of friends, heard Caruso singing *Pagliacci*.

He was unable, however, to place the device on general sale, because the Marconi Company, owners of the Fleming Valve patent, brought suit against him for infringing that patent and the case was decided in their favor. Nevertheless, his own patents, granted in 1907, prohibited the Marconi Company from making use of the third element. There matters stood, deadlocked,—the Marconi Company limited to the use of its not very sensitive two-element tube, De Forest unable to use his three-element tube for commercial radio purposes. . . .

Meantime the crystal detector arrived on the scene, and the quest for a good rectifier died down. The crystal was better than either the early Valve or the early Audion.

Both De Forest and Fleming believed the action of their tubes to be entirely due to the gas remaining in them, as was the general belief at that time. Though the tubes were called "vacuum tubes," it was believed that if they should be exhausted until their interior was in fact a perfect vacuum, the electron currents would then be unable to flow between the filament and the plate because they could not cross the vacuum gap.

It was whether or not this was true that came in

1911 to interest Dr. Langmuir, up in the laboratories of the General Electric Company and busy investigating all the different things that went on inside electric lamps. The particular thing that started his mind working on it was his observation of the result obtained when a very high current—two or three hundred volts—was passed through an ordinary tungsten lamp. These lamps were now being made with excellent vacuums; he noticed that the so-called Edison Current passing between the legs of the filament was very small. He had his assistant, then, deliberately introduce gas in different quantities into various lamps, and found that the more the gas, the greater the Edison Current. . . .

In August, 1912, he was still thinking about it. He made a special lamp with two filaments . . . and got results that indicated that there was some Edison Current even in a perfect vacuum. This seemed to challenge current scientific belief, and brought up the old question of whether or not Richardson's Law was true. Dr. Langmuir, who probably knew more about producing a perfect vacuum in a sealed glass body than any other man on earth at the time, decided to settle the question once and for all to his own satisfaction.

He made, then, three more lamps, each containing two separate tungsten filaments. While they were connected to the vacuum pump, being exhausted, he baked every bit of gas out of the glass walls by heating them for an hour or more in an electric oven to a temperature of 360 degrees Centigrade; to drive the gas from the filaments he heated those for thirty minutes to 2128 degrees Centigrade; he used a liquid-air trap, he used the "clean-up effect" and took every other precaution needed to get lamps that had a perfect, permanent vacuum of a very high order. This was in November, 1912.

Immediately on testing these "lamps" or "tubes" he discovered very interesting things. Thermionic currents *did* flow in vacuums; Richardson's Law was fundamentally correct, though conditions within the tube modified it at high filament temperatures. Best of all, this new vacuum tube would stand up under voltages as high as 250 volts between the hot cathode (the "filament") and the cold anode (the "plate"), whereas "gas" tubes like the De Forest Audion usually broke down and destroyed themselves when the voltage was raised above thirty or forty. . . .

The modern vacuum tube was born—a product of pure science in an industrial laboratory.

Dr. Langmuir went on studying thermionic emissions until he knew everything there was to know about them. During the remainder of 1912 and in 1913 he made dozens of these new "pure electron discharge tubes" in his laboratory. Always they acted the same, in contrast to the old gas-filled tubes which had been highly irregular in action. He measured their characteristics, worked out the laws of their behavior, and published, in the *Physical Review* of December, 1913, an article telling what he had learned.

The new principle found immediate application. Dr. Coolidge, who had invented drawn tungsten, was working to improve X-ray apparatus and tubes, hitherto erratic of performance and never entirely satisfactory; he applied Dr. Langmuir's thermionic discoveries and developed the modern X-ray machine, which did exactly what its user wished. Thus that aid to surgery was available when the war began. . . .

The development section of the laboratory also immediately applied the new principle to three-element radio tubes; tests and experiments showed what a contribution it was to the radio art . . . but General Electric was unable to commercialize the radio application—the basic patents were held by the Marconi Company and Dr. De Forest.

Nothing about the whole fascinating story of these tubes with their vacuums and flying electrons was more deeply interesting than was the final explanation of what had really been happening inside the old "gas" tubes.

Each tiny atom of gas present in the tube was, as Thomson had shown, made up of a positive nucleus surrounded by a number of negative electrons, the whole forming something similar to an infinitesimally small solar system, the nucleus being the "sun" and the electrons being the "earth" and other planets.

This atom was stable; the positive and negative forces in it were balanced, so that neither the negative filament nor the positive plate had any attraction for it, and it remained swimming in the space between the two.

When, however, the filament was heated, negatively charged electrons started boiling off from it, and darted toward the positive plate with velocities in the neighborhood of fifty thousand miles an hour! Inevitably, along the way some of them encountered gas atoms, the result being a violent collision, and inevitably some of these collisions had the following startling effect: they knocked an electron out of the atom, just as a giant meteor rushing through space might knock Mars, or Saturn, or the Earth out of the solar system!

Now that damaged atom had lost its balance. The freed electron, negatively charged, went rushing toward the plate. The remainder of the atom, called an *Ion,* being preponderantly positive, started on a mad journey of its own toward the negative filament, which it struck with such force that it knocked other new electrons out of it.

These new electrons, careening toward the plate with those being constantly boiled from the filament, collided in turn with new atoms of gas, and so the whole cycle continued! Therein lay precisely the reason why the De Forest three-element Audion had been so sensitive. Its electron activity was not confined to the electrons boiled out of the plate in accordance with Richardson's Law, but was greatly heightened by the presence of the gas. This was splendid in its way, but it prevented any such tube from ever acting twice in precisely the same manner—the amount of gas in it was constantly changing, and a very slight difference in the temperature of the filament made an enormous difference in the activity within the tube.

When more than about thirty volts was applied between the filament and plate of those tubes, the activity within them usually became so great that they destroyed themselves—literally the downfall of a universe!

Dr. Langmuir was the first but not the only man to discover the beneficent effects of a pure electron discharge.

In those same months of 1912 in which he was experimenting with his new theories and making such startling discoveries, the Research Laboratory of another great organization, the Western Electric Company, was engaged on a problem in land-wire telephony. The Western Electric Company was the manufacturing subsidiary of The American Telephone and Telegraph Company—parent of almost all the telephone companies in the United States—and this enterprise was trying to make it possible to talk by telephone from New York to San Francisco, a thing hitherto unaccomplished. There was to be a World's Fair in San Francisco in 1914 or 1915.

If it could be done, it was planned to open a transcontinental circuit during that Exposition.

So far, the distance between the two cities had been prohibitive. The fragile electric currents used to carry the human voice through a telephone wire were too weak to travel those thirty-four hundred miles; they dissipated themselves on the way, and though the wires existed that would connect the two cities, yet a voice could not be carried between them.

The problem was to find an "amplifier"—something that would replenish those fading currents at intervals along the route and give them new strength with which to continue their journey. There were many such devices in existence, but none of them that was suitable for telephony. The coherer, for instance, was an amplifier of damped radio waves—it permitted a battery to take up the work of operating the Morse Inker. The Morse "relay" was an amplifier commonly used in wire telegraphy. There were others, but the amplifier needed by the telephone company must carry on the voice waves without distortion. Its action must be electrical, not mechanical.

On October 30th, 1912, Dr. De Forest brought his Audion to the Telephone Company engineers, and they tested it for their purposes. The Audion was a radio detector, yet it had amplifying characteristics—a weak antenna current, led to the grid, controlled a much stronger electron current passing between the filament and the plate, so that the two fluctuated proportionally.

Because of its gas action, however, it distorted voice modulations except at very low powers—it could not handle sufficient power for the telephone job in hand. The engineers, who were able to pass one watt of electrical energy through a telephone transmitter, wanted their amplifier to produce at least that much—but the Audion would not yield more than about one one-hundredth of a watt. When it was forced higher it broke into "blue glow," indicating violent ionization of the gas, and ceased to amplify at all. Forced higher still, it destroyed itself. In any case its life was short—from ten to a hundred hours.

Dr. H. D. Arnold, in charge of the work for the Telephone Company, believed, however, that if he could pump a better vacuum into the Audion he could get more power out of it. Thus, starting from opposite premises, he and Dr. Langmuir worked toward the identical result, neither man knowing of the other's activity. Dr. Langmuir started to investigate phenomena occurring in vacuums; Dr. Arnold started to improve an already existing vacuum tube.

By the middle of 1913 the Telephone Company was sufficiently encouraged to warrant getting a license from Dr. De Forest to manufacture and use the Audion commercially. This was possible because the Marconi Company's Fleming Valve patent, which controlled the use of the Audion or any other electron tube, only covered the use of such devices for radio purposes.

Working ahead by the trial and error method, they had in 1914 produced tubes of the pure electron discharge type; had learned how to heat the glass and electrodes to get rid of any remaining gases, and were well on the way toward the modern device. Using their new tubes, they opened the New York-San Francisco telephone circuit in 1915 as they had hoped to; before long vacuum tubes were an integral part of the telephone system and had become so important that in 1917 the Telephone Company bought full ownership of Dr. De Forest's three-element Audion patent for the sum of $250,000, the inventor retaining only the right to use it for certain limited purposes, such as manufacture and sale to radio amateurs.

Dr. Langmuir applied for a patent for his high vacuum tube on October 14th, 1913. Two years later Dr. Arnold filed an application covering the same invention, and in the belief that he had started work on the Telephone Company's tube first, asked that the two applications be declared in interference with each other. The conflict between these two scientists gave rise to one of the bitterest contests ever waged in this country for possession of a patent, since each company knew what an important part the vacuum tube was destined to play in future radio and telephony. It was dropped during the war, while the two scientists were in government service, but reopened with fresh vigor immediately thereafter. Time and time again the issue was declared in favor of Langmuir, but Arnold and the Telephone Company appealed again and again, and carried it to ever higher tribunals. Nor for twelve years was the matter finally settled, when once and for all it was decided in favor of Langmuir. The patent was granted to him on October 29th, 1925, and became the property of the General Electric Company.*

These improved tubes immediately began to play a part in radiotelephony when, more or less simultaneously, three men discovered that tubes could be so connected as to become generators of continuous waves and used as a source of radio-frequency oscillations. Edwin H. Armstrong, a young man just out of Columbia University in New York and one of a new generation of radio engineers, had studied there with the eminent Serbian physicist, Dr. Michael Pupin. In 1913 he invented a radio circuit known as the "feed-back," by means of which a vacuum tube could be made to oscillate; it was patented

* Early in 1928, in a General Electric Company suit against the De Forest Radio Company, Judge H. M. Morris of the U. S. District Court for the District of Delaware, reversed these former rulings and held Arnold prior to Langmuir; as this note is written (September, 1928) the case has been appealed by General Electric to the U. S. Circuit Court of Appeals for the Third Circuit, and it would seem that this patent has again become involved in extensive litigation.

in October, 1914. Dr. De Forest and Dr. Langmuir devised circuits for the same purpose, but when the conflicts between the three were taken to court, the Armstrong patent was decided to be the predominant one.*

As yet vacuum tubes were too small, and even the new high vacuum tubes not far enough along to admit of this new continuous wave generator displacing either the "arc" or the alternator. It was immediately useful, however, for one thing.

The growing use of the "arc" had introduced a new problem in radio receiving. The continuous waves, coming through the ether and rectified in the ordinary way—as by a crystal detector or a vacuum tube—were inaudible when they reached the telephone receiver because their frequency was too high to register on the human ear. Dr. Fessenden, in the old days at Brant Rock, had come up against the same problem in his early use of the "arc," and had proposed as a solution the "heterodyne" principle. If, for example, continuous waves were coming into the antenna with a frequency of 300,000 cycles per second, he proposed generating as an integral part of the receiving apparatus, similar continuous waves of some such frequency as 301,000 cycles per second,

*** The "feed-back" patent became, like that of the pure electron discharge tube, the subject of a *cause célèbre;* it remained in the courts for years. In October, 1927, the Circuit Court of Appeals for the Third Circuit, reversing former decisions, held that De Forest's "tube-oscillator" had priority over Armstrong's "feed-back," whereupon the case was taken to the United States Supreme Court. It has not, as this is written, been finally adjudicated.**

and combining—mixing—the two together; the result would be a "beat" note of 1000 cycles frequency, which would be audible in the phones. Anyone who has ever heard a twin-motored airplane has heard the throbbing "beat" of the engines as their exhausts first coincide and then draw apart again, repeating the process over and over. Anyone who has ever heard a radio loudspeaker shrill and whistle has heard "heterodyning" as two continuous Hertz waves of nearly the same frequency merge within the set.

Fessenden had proposed the use of a small "arc" to produce the heterodyne. That had been clumsy and unsatisfactory, but now, with vacuum tubes and Armstrong's oscillating circuit, the principle could be used and began to find its way into receiving apparatus. A little later Armstrong himself combined the "feed-back" and the heterodyne in a particularly efficient way, which came to be known as the "super-heterodyne." . . . He owned the "feed-back" patents; those covering the heterodyne were the property of the National Electric Signalling Company.

The vacuum tube amplifier at last offered a satisfactory way of impressing the human voice upon a continuous Hertz wave; in Schenectady they telephoned back and forth to Pittsfield, Massachusetts, where the General Electric Company had another factory.

The Telephone Company had been interested in radiotelephony for years. Some of their telephone circuits crossed lakes and other large bodies of water, and they had always had trouble in using underwater cables—cables distorted the voice to the point of unintelligibility. As early as 1909 they had investigated radio as a substitute; they had had men at a Fessenden Brant Rock demonstration when he succeeded in telephoning by radio between that station and Washington. Indeed, after some months of observation and experimenting, they had almost bought his system—but a sudden change in Telephone Company management had brought an end to the negotiations.

Now, with the tube to work with, they took the matter up again. In 1915 they built a giant experimental transmitter and installed it, with the Navy's permission, at Arlington. They were still far from a tube large enough to handle any great amount of power, but they connected five hundred of their small tubes together in a huge "bank," generated continuous waves with them, and used others to add the voice modulation. Speech from Arlington, then, was heard in Paris. It was heard in Honolulu. A naval officer on the Pacific Coast sat at his desk in the Mare Island Navy Yard and spoke into his office telephone; his words travelled across the country by wire to Arlington, where they were launched into space and to a battleship in the Atlantic. The battleship's captain answered him, likewise by telephony. Men were beginning to achieve what had before been only a dream.

And at Schenectady, in 1915, the alternator that had so long been such a goal for radio engineers was finally perfected. Ever since the failure of the early machines, the General Electric Research Laboratories had been digging away at the problem; in 1915 Dr. E. W. Alexanderson, to whom it had been specifically assigned, completed the work begun so long before, and the Alexanderson Alternator, a masterpiece of electrical machinery, was ready to go to work.

Not only had the alternator itself been made practical, but all the other things necessary to complete a "system" of radiotelegraphic transmission—delicate devices for controlling the speed of the spinning giant, means for putting its great power into the antenna, a new antenna far better than any before. . . .

Fessenden's dream of 1901 was brought to realization, but by other men and only after the expenditure of thousands of hours of organized creative effort, and of enormous sums of money.

In this, in vacuum tubes, in radio circuits, the laboratories had accomplished what the early inventors had failed to do. It was an indication of the passing of an era. The solitary scientist, snatching time from classroom work to experiment in a university laboratory, was no longer to be the leader of the world's scientific progress. His place was taken by his modern brother in the industrial research laboratory. . . . No longer was the "inventor," the self-educated man with a dream, to be the sole producer of new commercial applications—his place was taken by the modern engineer.

CHAPTER SEVEN

WAR

SINCE the outbreak of hostilities in 1914 only rumors and propaganda had reached this country as to the actual state of military activities—and all activities were military—on the part of the European combatants. Censorship had been rigid on both sides; the official *communiques* had contained only what their sponsors wished to reveal to the world, and though by piecing together all the information available it was possible to reach a general idea of the war's progress, the means by which results had been obtained were very successfully kept secret.

Only in 1917, when German submarine activities and the devastating scope of the conflict finally drew the United States into the great clash of economic forces and rivalries, did our military people at last become "insiders" entitled to know the facts. Then the story of the part radio had played in the war came out.

Germany, in 1914, had been as well advanced in the art, technically, as any other nation in the world. She had not done so well in its commercial exploitation, largely due to the successful competitive activity of the associated Marconi companies, but in its military use she had kept abreast the times, as she had in the military use of so many of the era's novelties. Her Navy was well equipped, well trained in radio-communications; it had ample shore stations. Furthermore, realizing the likelihood of losing her cables she had erected the high-power Nauen station and was working on another one at Eilvese, near Hanover, to tie her to her colonies in South Africa and the Pacific, and to the outside world.

France, with less occasion for the use of the art, had done less toward adapting it to the military, though her Navy was equipped. Britain, on the other hand, had always been radio's most active proponent. Her Navy was as well outfitted and trained as Germany's, if not better. She had had, moreover, other extensive government radio activity—the Post Office had been operating the coastal stations—and she had reserves of operators and designers, thanks to the extensive operations of the Marconi International Marine and the Marconi factories that had been building apparatus for the All-Red Chain and the Marconi high-power stations in the United States, Honolulu and Norway.

The military use of radio had been foreseen since the very introduction of the art. Governments had always been radio's best customers. And from the August days in 1914 when the entire Second Western Civilization became enmeshed in an orgy of bloodshed and wholesale self-destruction, both the Allies and the Central Powers had carried on radio activities along lines foretold years before in some strangely prophetic words.

In 1899 Sir Oliver Lodge, then deep in "syntony," had said in an article for the *Anglo-Saxon Review*: "There is also about this method of communication the usual objection to combat as regards secrecy. Since the sound spreads more or less in all directions, it can to a great extent be heard by other persons; . . . in war times eavesdropping undoubtedly may exist, and with it the two other hostile operations, drowning of speech on the one hand and communication of lying messages on the other, as they will doubtless hereafter occur in more modern methods of wireless telegraphy."

Fifteen years later, here was the realization of Sir Oliver's prediction, all except the "drowning of speech"—for that bugaboo of many years standing—intentional interference—was strangely absent from wartime radio activities. It was too important to hear what the other chap was saying!

Britain, of course, promptly *did* cut all cables leading from Germany, and put a rigid censorship upon all others that might conceivably be used, directly or indirectly, for German messages. All her commercial radio was taken over by the military . . . then was it seen in the United States what it meant to dominate world telegraphic communication—one could scarcely send a cable from this country without a British yea or nay. . . .

And Germany promptly fell back upon radio. The ether of the Channel was full of the smooth, efficient wireless talk of her Fleet as her ships and their bases exchanged messages in intricate cipher, hurrying as if to get them through before "interference" came. It never came; she was granted the utmost freedom to use her radio all she wished—and she took the bait.

For the British were pursuing a policy that it was more blessed to receive than to give. While they themselves maintained "wireless silence" except in emergencies, they listened, ears glued to receivers, "eavesdropping" for all they were worth. Every recorded message was telegraphed by land-wire to the British Intelligence Headquarters at the Admiralty, and there, day by day, the whole was coördinated, pieced together—a continuous record of the Teuton's radio-communications.

At the outset this meant little more than the arduous collection of meaningless hieroglyphics which the British cipher experts tried rather vainly to unscramble, for they had neither the German code books nor their coding machines. But that autumn a German cruiser was sunk during a sea-brush with the Russians; divers went down after the vessel's safe, recovered it, and the Allies came into possession of a set of the latest German codes. And in January, 1915, the surf at Yarmouth cast a derelict German submarine upon the shore; all its crew were found to be mysteriously dead although everything was in perfect order—the Admiralty was grateful for another set of confidential pamphlets, delivered to hand. . . .

With so much to go on, the "eavesdropping" system became of inestimable value. To supplement the contents of the messages themselves, "direction finding"—the determination of the point from which radio signals originated—was highly developed. The direction finder, or "radio compass," had been the 1907 invention of two Italians, Bellini and Tosi, and rights to use it had been acquired in 1912 by the British Marconi Company. Now the instrument—based on the "loop antenna" principle—was much improved and each of the coastal stations equipped with it, so that with the report of each intercepted message there was included the direction from which the signals had come.

London was able to "plot," by radio cross-bearings, the movements of every German ship operating in the war zone! For the Germans continued, obtusely, to place the utmost confidence in their ciphers and to use the radio most methodically and efficiently. The codes were changed at intervals, but even that situation Britain kept abreast of—her warships were regularly sinking submarines, and she had a special "mystery" ship that was immediately dispatched to the scene of every such victory. Working in secret, it located the hull and put down deep-sea divers who walked in among the dead men . . . and out again with the precious little tin dispatch box, always found in the same place, containing the codes. At least once a month the Admiralty's information was thus refreshed, and the Germans never knew . . . until long after the war was over *no one* but a scant handful of British Intelligence men ever knew how those codes were gotten.

After 1916, however, Germany did grow wary of the direction finder and cut down on the use of radio at sea. Her submarines on long cruises into the Atlantic had to report nightly, but her ships in the North Sea were as silent as England's. She, too, established a direction-finder listening service . . . and sometimes those listening men on both sides of the Channel felt chills go up and down their spines as cruising men-o'-war sighted the enemy and the silence was broken. The ether fairly throbbed, then, with full-power calls for help, orders to supporting ships, the roar of interference . . . once it was the Battle of Jutland, the Grand Fleet versus *die Hoch See Flotte,* and they could try their utmost to piece the fragmentary bits of messages together, straining to learn how the tide of sea-power was going—but unsuccessfully. In naval action, interference strangled radio, as men had always feared it would; if the British had been able to get wireless reports through, that battle might have gone differently.

In the combatant armies radio found its greatest use in air activities; it added a voice to the eyes furnished by reconnaisance and artillery observation planes, and made surprise movements on the part of either side more difficult than ever. It was used to some extent, too, for communication between front lines and base headquarters, though the wire-telegraph and the "runner" handled the bulk of that work.

And it was used for propaganda. Deep within Germany the station at Nauen broadcast "true" reports of the progress of military events and of affairs in Germany to anyone who would listen, and by that medium the German side of things was gotten to the outside world. Much has been made of this propaganda, but as a matter of fact it was in no wise different from that issuing from all the combatant nations. Britain's, and that of France, did not come by radio for the simple reason that the cables and mails carried their traffic, but like Germany's their *communiques* suppressed the unpleasant, stressed what it was desired to stress, and invented much that would sound plausibly meritorious when consumed by those whose good opinion and support it was so to their advantage to maintain. Propaganda was as much a weapon as the bayonet, and a little later we were to make full use of it ourselves . . . not only by radio but for home consumption. That which emanated from Nauen was not as effective as it might have been; it lacked subtlety, which was a fatal defect.

But there were mistakes more serious made in Nauen's use. Germany's reliance on her ciphers was bad enough at sea, but when she used those same codes for her Nauen messages to her agents abroad she laid herself wide open to the thrusts of her enemies. For example, the revelation made to the United States by Britain, that the German Embassy in Washington was actively striving to bring Mexico—with whom we were having a private squabble—into the war on the side of the Central Powers, stirred up a fine hornet's nest and took us a long stride on the way to joining the ranks of the Teuton's antagonists. Britain, of course, conveyed the information for purely altruistic purposes . . . having had it handed to her on a silver platter by a Nauen message which she decoded half an hour after it was intercepted. . . .

Perhaps it is safe to say that the use of Nauen did as much to aid Germany's enemies as to further her own ends. At any rate, this, and all her other radio activity, was marked by the same blind faith that she placed in all her instruments of war . . . that they could not be fallible. It may even be that she intended that some of those messages be intercepted; but if the error was psychological, it was none the less an error.

The radio problem of the United States, upon our entering the war, was one of an entirely different order than that of any of our allies. Once committed to the combat, our war effort took three main channels, all of them gigantic undertakings and all of them, by our very geographical location, involving ocean transport and therefore the use of radio on a scale hitherto unprecedented.

In the first place, we threw behind our weary copartners our full support in the matter of munitions, supplies and food; thousands of tons of those commodities had to be provided, assembled and shipped to specific destinations on the other side of the Atlantic. Secondly, we embarked upon the training for combat of several millions of our population, their transportation overseas to the area of hostilities, and their maintenance there—which meant the sending across of an additional stream of ships with munitions, supplies and food for their exclusive use. Lastly, we threw our not inconsiderable naval strength into active participation in the British Navy's task—that of keeping command of the seven seas with their vital trade routes, and particularly of the sea approaches to France. This was active naval warfare, for while Britain had declared a blockade of Germany, Germany had likewise given notice of a submarine blockade of the war area and was making vast strides toward the carrying out of her threat—in those spring and summer months of 1917 she was sinking a million tons of allied shipping every month, and no ship at sea, whether merchantman or man-o'-war, was free from the momentary possibility of being torpedoed.

For the seaborne tonnage immediately necessary, it was fortunately possible to depend upon Britain, for we had only a fraction of the needed amount; British shipping carried most of the food and men across in the first months. But ours was effort planned on a greater scale than a matter of months; since it was so vital that there be a wide, strong bridge of ships across the Atlantic, we determined to have ships of our own—we were again to become a great maritime nation, if only for the duration of the war. . . . The Government, through the Shipping Board, bought ships wherever it could lay its hands on them. In newly erected shipyards bottoms were put down by the hundred, wooden bottoms, steel bottoms, fabricated hulls, destroyers, "Eagle boats," fat-bellied freighters, submarine chasers, every sort of ship that sailed the seas. . . .

And every one of them had to be equipped with radio.

We were to fight in France. We were to set up an American community of fighting men there, and its roots were to be in the United States. Did it need shoes, or guns, or shells, they not only had to be brought from America, but word had to be gotten to America that they were needed. With every shipload of men that went over, the transoceanic communications problem increased.

To expand the facilities? . . . Radio.

We were to fight at sea. Destroyers were to rendezvous with convoys, to circle and dart after threatening submarines, to spend long days searching the sea lanes; battleships were to stand out from port cleared for action; far-flung ships to be manœuvred as one. . . .

Radio.

At the outset there had been those two radio structures, commercial and naval, one serving the ends of maritime commerce, the other those of our arms at sea. And just as military need came entering into every man's business until all commercial enterprise from corner grocery stores up had its military aspect, so the war now eliminated commercial radio-communications entirely. The Government commandeered the fifty-nine coastal stations that had served those ends; by Presidential Proclamation they were handed over to the Navy. All others—amateur and privately owned—were closed, for there must be no instruments that might communicate with the enemy or his ships.

In accordance with the plans worked out back in 1914 when there had first begun to be a European war and a possibility of our being drawn into it, radio communications in and for the United States became the province of the Naval Communications Service. It was a bigger job than Naval radio men had ever handled before; too big, indeed, for them to handle alone,—they must have the support of industry and men. It divided itself into three channels: first, all government ships—men-o'-war, transports, and Shipping Board freighters—must be radio-equipped and provided with operators; second, the chain of shore stations, both naval and commercial, must be operated to serve all shipping, government, private and foreign; third, overseas radio-communications with Europe and with our colonies must be made capable of replacing the cables if those were cut, and in any event, of handling at least a third of the Atlantic telegraphy.

This meant expansion to the utmost. The great increase in shipping and the need for keeping it moving in all weathers meant that even with the joint use of naval and commercial stations the present facilities would not serve. The immediate need was for operators—there were only 1031 in the Navy when war was declared—and instruments. The call went forth for both of them.

It found the radio industry in a singular situation. The period since 1912 had been one in which the American Marconi Company had, through its fundamental patents and business activity, become more firmly entrenched than ever as dominating both the manufacturing and commercial communications fields. The Marconi and Lodge "spark" and "tuning" patents had been repeatedly held by the courts to cover all "spark" radio; the Fleming Valve patent controlled the radio use of vacuum tubes. No one could legally manufacture or operate these devices without infringement, and every company that had tried to had been very legitimately sued. The Marconi Company had a factory at Aldene, New Jersey, in which was manufactured about ninety-five per cent of all the radio apparatus made in the country; their ship and shore stations handled about the same proportion of the commercial radio traffic.

At the same time there were a number of small radio manufacturers hanging on, chiefly by virtue of occasional Navy orders and because they had something valuable enough to trade for a limited right to use the Marconi patents. There was United Fruit's Wireless Specialty Apparatus Company, for example, which owned the crystal detector patents; it was in effect subsidized by its parent and by the Navy. There was the National Electric Signalling Company, still being operated by a receiver; in 1914 it had responded to a Marconi suit by bringing a counter-action of its own charging infringement of the Fessenden high-frequency "spark" patent—with a resultant withdrawal of both suits and an agreement between the two companies. For two years it had gotten a twenty per cent royalty on all Marconi apparatus . . . but just latterly the Fessenden patent had lost out in another suit brought against another party, and the Marconi Company had notified the National Electric Signalling Company that it was terminating the agreement. . . .

De Forest was completely tied up, though he had just sold rights in the Audion to the Telephone Company for $250,000. There were two infringing tube manufacturers—Moorhead and Cunningham—out in San Francisco; Moorhead was selling to the British government and to amateurs, Cunningham by mail-order to amateurs. There were perhaps a dozen small concerns making radio parts—condensers, coils and the like—for the amateur trade.

Even the two companies which operated independently of Marconi patents were very limited by them. The Federal Telegraph Company, manufacturing the "arc," was free as far as that instrument was concerned, but it had been in difficulties when, desiring to make a smaller apparatus for use on ships, it had gotten mixed up with Marconi and been forced to stop. The other, Kilbourne and Clark—in Seattle—had been in the radio manufacturing game for several years with some "spark" sets held by the courts to operate on a different principle than those of Marconi—due to its litigation with Marconi, however, Kilbourne and Clark was not flourishing. An electrical house, it had first made motor-generators for United Wireless in 1911. After United's demise there had grown some demand on the Pacific Coast for sets to be owned by ships—rather than leased—and in 1914 it had started making "Simpson" and "Thompson" sets for that purpose. In 1915 the Marconi Company had brought suit, alleging infringement; the case had dragged on, and although Kilbourne and Clark had finally, at the time of our entry into the war, been awarded the decision, Marconi had appealed. The Seattle company's business had fallen off to nothing, due to the reluctance of shipping people to purchase apparatus which they might later be enjoined from using if the Marconi Company eventually won out in the litigation. A Kilbourne and Clark subsidiary—the Ship Owner's Radio Service Company—formed to care for their sets installed in ships, was likewise a dead issue.

This was the industry that was available to take care of the nation's giant radio expansion!

The Marconi Company, its strong man, was alive to the hour. For all its British control, its personnel was entirely American and as alert to the national task as any body of men in the country. Its President was John W. Griggs, former Governor of the state of New Jersey; it had a splendid record of service and a fine *esprit de corps*—Marconi operators "got their messages through," and if those happened to be SOS calls, stuck to the key until help came or the ship took her plunge . . . they had a tradition . . .

In February, 1917, when the situation with Germany became really acute after that nation's pronunciamento allowing us one ship a week to England—all others to be subject to the torpedo—the

Marconi Vice President and General Manager, E. J. Nally, had written to President Wilson offering all the company's resources and men to the service of the nation. Now, as war was declared, they were called on. The factory went to work full blast; the forty-five coastal and eight high-power stations became Navy stations, and 456 Marconi men turned to Naval service. . . .

But just as the job in hand was bigger than the Navy, it was bigger than the Marconi Company—"spark" apparatus was not enough. The job was bigger than the Federal Company—the "arc" was not enough. These were times when America was calling on her foremost internal-combustion engineers to develop a "Liberty" airplane motor, on her shipbuilders to erect a Hog Island. The Navy turned to the General Electric Company, to Westinghouse, to Western Electric, foremost electricians in the country, if not in the world, for a new radio.

All of these companies had done radio research in their laboratories. None of them had been able to enter the radio field because of the patent situation, nor had they cared to. Now with one stroke the Navy cut the Gordian knot; it assumed full liability for any patent infringements caused by manufacturing Navy radio apparatus, and the full power of America's electrical engineering strength was free to get behind the radio job.

The result was a new radio industry, and it made the new radio that was needed.

General Electric at Schenectady and Westinghouse at East Pittsburgh made the wonderful new vacuum tubes, and receiving sets to operate with them as no receiving sets had ever operated before; they made transmitters, too. Western Electric made tubes and radio telephone apparatus—hundreds of sets went into submarine chasers, destroyers, battleships. And the little companies—Kilbourne and Clark expanded; the National Electric Signalling Company was reorganized and allowed by the receiver to sell its assets to the International Radio Telegraph Company, a new concern with the same old owners, which proceeded to get into full swing on government contracts; the De Forest plant blossomed into life; Federal concentrated on bigger and bigger "arc." Marconi did yeoman service. Half a dozen new organizations came into being. Freed from patent restrictions and with almost unlimited millions being poured out of government coffers for the apparatus that was now as precious as jewels, all these suddenly created radio factories ran full blast.

It was greatly to the credit of the Navy that of all the flood of dollars poured out by our Government for the prosecution of the war, that part which was expended by the Bureau of Steam Engineering for the Naval Communications Service was administered with considerable wisdom. It was, of course, impossible to foretell the part that radio would play in the world's affairs when the war was over. Radio seemed then to have but two definite applications, overseas communication and communication between ships and shore, and there seemed every reason to believe that the future would bring merely higher developments along those lines. At any rate, the Navy decided to build a structure that would serve not only for the war job, but for the peace task that would come afterwards. This decision was stimulated by a hope that after the war at least the ship-shore end of radio-communications would remain under naval administration.

There had been thirty-five Navy coastal stations when the war started; added to these were the forty-five Marconi ship-service stations taken over for government use on compulsory "duration of the war" leases. To make the system the last word in completion, the Navy now set about building thirty-nine more, making a total of 119 stations for communicating with ships. . . .

In 1918 the forty-five Marconi stations were purchased outright. It came about in a rather unusual way. . . . The Navy, be it understood, was charged with providing radio service for Shipping Board ships as one of its jobs. 335 of these ships, when taken over by the Board, were equipped with rented Marconi apparatus and using Marconi service; for a time the Navy continued this arrangement and paid the rentals. It was then decided that the apparatus ought to be Navy property and the Marconi Company was asked to name a price for it, to which the company rather indignantly replied that such a sale would take away the major part of its business—it had only 500 ship installations in all. But the Navy persisted in its stand and the Marconi Company then said that if the ship stations must be sold it would like to get rid of the coastal stations too, for the one was no good without the other . . . and as a result the Navy bought the whole lot for $1,450,000 and the Marconi Company was virtually out of the ship-shore communications business. Its eight high-power stations were still its property, however, and it had its factory. . . .

The new Navy chain of coastal stations was operated in a way much improved over anything previous. Hitherto coastal stations had been single units, each one distinct. Now they were grouped, and use was made of remote control. For example, the Third Naval District, extending along the Atlantic coast from Rhode Island to Barnegat Light, New Jersey, had eight coastal radio stations. These were all controlled by land-wire from the office of the District Communication Superintendent at 44 Whitehall Street, New York City, where all the actual sending and receiving was done; the men at the stations themselves merely kept the apparatus running. This meant that stations whose masts, etc., were as much as two hundred miles apart were operated from a single room where they could be perfectly supervised and controlled.

In those days when ships were so precious it was vital that they be about their missions with as little delay as possible. Fog, then, was as much an enemy as the gentleman in the coal-scuttle helmet, and to overcome it that same radio compass that had been so useful to the British in fixing the position of submarines, was employed, so that ships, groping their way through fog, might be told by radio where they were. In 1918, using an improved version of the compass (developed three years earlier by Dr. Kolster of the Bureau of Standards), eleven radio compass stations were installed and thirteen more begun. . . .

And at sea, during those war days, the Naval Communications Service and the radio manufacturers put 1500 new radio installations in Shipping Board ships, and over 1200 in naval vessels!

These many radio sets had to be manned. At the outbreak of the war there had been only 1031 radio operators in the Navy; the Marconi Company contributed some 450 more, and many of the nation's 5000 radio amateurs brought forward their specialized knowledge and offered it as their "bit." But operators were needed in numbers larger than these. In June, 1917, a training school was started at Harvard University; 350 men formed the first unit. When the Armistice was signed 7000 operators had been ground out by quantity production methods; there were 3400 under instruction, and new ones were being sent to sea at the rate of 200 a week. . . .

In such wise did America prosecute the war!

Communication with France. In October, 1917, there was an inter-allied conference of military and naval commanders at which was discussed the possibility of the Atlantic cables being cut. This was certainly a thing to consider, for the U-boat was moving about with a great degree of impunity. . . . The Conference decided that trans-Atlantic radio service should be improved until it was capable of replacing the cables. . . .

Good old quest! Replace the cables! At the time there were those same Marconi, Telefunken and Arlington stations along the coast, that had been batting their heads against that problem for so many years. But now the sea *must* be conquered; this was more than a question of revenue.

The latest high-power transmitting equipment in any of those stations was the big "Arc" that the Navy had put in at Sayville, and that was not adequate. The latest receiving equipment was that in the Marconi receiving stations at Chatham and Belmar, and it was not adequate. No, a fresh start must be made; those media that had been so eagerly sought after for all these years must now be produced.

There was, in the Marconi station at New Brunswick, New Jersey, the first of the Alexanderson Alternators, a fifty kilowatt experimental machine placed there by General Electric before the war for demonstration purposes. It was admittedly the most efficient transmitting instrument yet produced. In the works of the General Electric Company at Schenectady there was another of those Alternators —a big one, two hundred kilowatts—which had been made as a sample of what the final product was to be. It was worth $137,000. The Navy went to the Marconi Company and asked them if they would buy it and install it in the New Brunswick station which the Navy was now using . . . and the Marconi Company refused. At that juncture General Electric itself volunteered to put the machine in, and so in it went, a huge, beautiful humming generator with great control boards and a long Alexanderson antenna stretching away toward Europe. The installation was complete early in 1918, and from thenceforward the problem of getting radio messages to Europe was solved. . . . The Atlantic had found its master.

The return route was more difficult. There were only three Allied high-power stations that would in any way serve—those at Rome, Lyons and Carnarvon—and none of them was fully adequate. Lyons was decided on, improved, and used . . . thanks to the revolutionary new vacuum tube receiving equipment in the United States. For a modern receiving station was being built and the Navy, the scientists and the manufacturers were working together to make it the last word in the art's development.

The site chosen was at Otter Cliffs, near Bar Harbor on Mount Desert Island, Maine—a location of singularly good radio receptive qualities. The apparatus made anything used hitherto seem old and antiquated; it had high-speed photographic recording instruments which would hum in response to automatic sending—a hundred and fifty, two hundred words a minute could be caught as they flickered across the sea. . . .

Another huge radio station was installed at Annapolis, Maryland, across the Severn River from the Naval Academy; this made a unit with Arlington. The station at Bar Harbor made one with New Brunswick. . . .

That was two "cable" circuits. . . .

And for better sending from the European end, the largest station in the world was built. At Bordeaux, France, for the French Government, the Navy, the Army and the American manufacturers, all working together, put up eight 820-foot masts supporting an antenna nearly a mile long; it was equipped with a 1000-kilowatt Federal-Poulsen "arc" transmitter that was to send out ether waves nearly ten miles from crest to crest—the Navy felt that in spite of the wonders of the Alexanderson Alternator, this giant, five times the Alternator in size, would be a final, perfect guarantee of getting messages across.

This station, known as the "Lafayette," was the

high point of the twenty-year effort to conquer the Atlantic by sheer size and power. It was not completed until after the Armistice; and a proposed sister to it in Monroe, North Carolina, was never started . . . but when finally commissioned the Lafayette station was a monument to the magnitude of an effort. Man, in striving to make sure that he would cross the sea with his electric words, made too sure, and the station was more powerful than necessary. . . .

But the Atlantic was very definitely conquered in 1918. The long lists of casualties that appeared that year in the newspapers were carried across from France by radio. The cables might have been cut, then, without our suffering impossible handicaps. Full transoceanic success was reached about a year after we entered the war; during that same year we had trained soldiers that were now landing in France a million strong every month. The whole effort was one so huge, one of such entire concentration and coöperation on the part of this nation that it is a question whether anyone stopped to realize fully just how great was the might and power of America as she flexed and bulged her muscles. And she had not even been pressed. . . .

The strictly military uses of radio followed the lines previously worked out and established by the combatants. The Army used it for reconnaisance and artillery observation aircraft, and to some extent for communication between the fighting lines and the various base headquarters. Those ships of the Navy that went to the war zone to join the British Fleet had their radio equipment altered to conform to that service; their operators learned and followed out British procedure.

One strange phenomenon of the war use of the radio art by navies was the growth of the "broadcast." Radio's ability to broadcast the messages of one station to many listeners had been a great reason for its compulsory adoption by ships. Now, again, the broadcast came to the fore. Radio's weakness—the thing that had caused Britain to maintain "wireless silence"—was the fatal fact that the position of an answering ship could be found by radio compasses. That weakness, it came to be realized, could be overcome if ships did not answer, and so at certain hours daily the sending stations simply broadcast all their messages for ships and expected no reply; by repeating each message in two or three successive broadcasts it could be made certain that it would get to its destination. . . .

On November 11th, 1918, the United States Naval Communication Service was an establishment to be proud of. . . .

CHAPTER EIGHT

AND WHAT CAME AFTER IT

THE western world had spent two decades in moving toward war; eventually it had been given over to the hands and ways of military men, and the conflagration had spread until all lands, all peoples were drawn into it. Generals, Admirals, had been more powerful, more looked to, then, than captains of industry. The whole world learned the ways of soldiery.

It would be unworthy of the military men to say that they hoped such a condition would continue indefinitely. No, they wanted the war to end just as badly as anyone else did . . . but being human, they were fond of their power; they believed in the beneficence of their guidance in this hour of need, and unquestionably they were the least bit shocked to see how violent was the world's repudiation of all things military in the weeks after November 11th, 1918.

The Armistice produced a let-down of tension as sudden and as sweeping as had been the patriotic emotion that drew all persons into the war effort when war was declared. The same men who had clamored to be allowed to get to France, to leave their civil occupations and the routine of peace, could not rest, now, until they could pick up the old ways. Youth might be uneasy; young soldiers might want to drift after the horror and license, but there were hundreds of thousands of others with families, women, children, with life-tasks, life-responsibilities waiting. There were latent careers a-borning. There was living, and growth, ahead. And the civil population suddenly realized that it was sick of war.

What a pet word *Normalcy* was!

Yes, the world was going back to the ways of peace as fast as it could get there, and there were Generals and Admirals who had been running things and now found that things had stepped right out from under their hands. . . .

Armistice.

The Naval Communications Service had had a dream for many years; a dream of a smooth, efficient radio service for the United States. Government monopoly. Naval operation. In a tidy, military manner, with proper precedence for naval ships and government business . . . it was a dream born back in the very first days of radio—engendered then by antagonism to the Marconi Company —and kept alive, constantly growing stronger, since then, by a sense of national rivalry toward that enterprise with its British affiliations; after 1914 they had come to think of Marconi radio as British government radio. . . .

So they had built toward their dream, worked toward it; it had prepared them to run the radio when things were given over to the military to run and the world went warlike. As early as 1917 they had gone to Congress with a bill providing for Government radio monopoly; in January of that year there had been hearings before the House of Representatives Committee on Merchant Marine and Fisheries—in charge of radio—but the Bill had been tabled because of the old chronic objection, among our legislators, to Government Ownership in any form; it sounded socialistic, or militaristic, or something . . . and besides, there had been certain sad experiences along those lines. . . .

But the Navy had worked ahead anyway, and during the war had actually achieved ownership of almost every radio facility in the United States by the simple expedient of buying up stations. And by the use of the money at their command, they had undertaken the active stewardship of radio's future. For example, in 1917 the Marconi system had moved to extend its service to South America by means of a proposed high-power station on the Gulf of Mexico. For this purpose it had formed the Pan-American Wireless Telegraph Company; and so that it might use the "arc," it had admitted the Federal Telegraph Company to a one-fifth share in the project—the remainder being two-fifths British Marconi and two-fifths American Marconi. Such plans for post-war penetration had alarmed the Navy; they read into them, moreover, machinations for the control of the "arc," and to preserve the sanctity of that American apparatus, bought from the Federal Company the patents covering it! The cash involved in this transaction, which also included the Federal high-power stations, was one million dollars. . . .

On November 11th, 1918, then, the only commercially owned radio stations of any importance left in the United States were the eight Marconi high-power stations, two in Massachusetts, two in New Jersey, two in California and two in Hawaii, and the one "French" station at Tuckerton—the Navy had bought Sayville, along with the patents of the Atlantic Communications Company, from the Alien Property Custodian. All these stations the Navy was still operating, under compulsory leases. . . . It seemed that the matter of the future was well in hand; that inevitably this nationalistic government control would continue when peace came. To cinch the question, another bill was gotten before the Congress.

Hearings on this one were held in the middle of December, 1918—but the war was over. . . . There was persistent opposition, not only from Congressmen, but from men who wanted again to make money out of radio, men who had developed it, put money into it,—from the Marconi Company, and from others.

Whether or not the military men realized it, and no matter how carefully they had planned, war and military control were things of the past, now. The bill was tabled, like the other—and there died the last chance for a government monopoly of radio in the United States, for the commercial interests were scrambling to get back into the field, and once they had a toehold the spell would be broken. . . .

Commercial radio, however, both in its manufacturing and operating branches, was suffering most acute pains as it cast its eye over the situation and pondered upon what to do. All the manufacturers, large and small, who had built big plants to fill war orders, had just suffered the shock of having those orders cease abruptly . . . and now they enjoyed an added jolt as the Navy announced that it could no longer undertake to shoulder the responsibility for patent infringements. The open purse of war was closing; the great inflation was over. Purely the creation of military exigency, the artificial radio industry was brought to earth with a thump. Only those holding basic patents could survive . . .

and even the basic patent situation was completely upside down as the result of the new radio developed by the war.

To General Electric, Westinghouse, and Western Electric, who were not primarily radio organizations, the edict of course simply meant closing the radio shops—but those like Kilbourne and Clark were in a genuine predicament. That company could not survive the reconstruction period; it hung on for a while, but in 1920 its management passed to a committee of its creditors. De Forest, and likewise the International Radio Telegraph Company (which had been the old National Electric Signalling Company and had the Fessenden patents) turned to making all sorts of electrical odds and ends—radio was out of the picture. The Moorhead Laboratories and Cunningham, makers of tubes, made quasi-legal arrangements with Marconi . . . but there was handwriting on the wall. A man named E. J. Simon, who had formerly been with De Forest and had made a fortune out of a shoestring on war orders, went off to Europe to see if he could garner in any new patents there. . . .

The Communications Companies were little better off. Federal, when the Navy had taken over its Pacific Coast stations, had leased land-wires from The American Telephone and Telegraph Company and carried on its business between San Diego, Los Angeles, San Francisco, Portland and Seattle by those means; it now prepared to build radio stations again, and bought its "arc" patents back from the Navy (which retained the right to manufacture and use them)—it faced heavy investments.

And the American Marconi Company . . . gazed about at the wreck of a business. What were *they* to do? Erect forty-five new coastal stations to replace those the Navy had bought? Was there enough money in it to make that worth while? There would be, of course, the ship-rental service—that was bound to continue because of the 1910 law . . . but the 330 stations on Shipping Board ships, sold to the Navy . . . the Navy was casting off Shipping Board radio, now; it made a situation not without its element of humor, for the Shipping Board was inquiring whether the Marconi Company would not buy some of those ship stations back again . . . the ships were being laid up to rot . . . the Marconi Company did so. . . .

To complicate matters there was a recently inaugurated edict that no company could manufacture radio apparatus for the Government unless it could swear to 51 per cent American ownership, and though 21,664 of the 23,027 Marconi stockholders were residents of the United States, the large blocks of stock were owned abroad; the company was unable to furnish the necessary affidavit. . . .

But the situation was not altogether bleak. American Marconi was strong financially, and there were the ample reserves and support of the British Company. The big post-war Marconi job, it was decided, would be trans-Atlantic radio. Let the Navy keep the coastal stations.

For transoceanic radio was now entirely feasible, thanks to the Alexanderson Alternator. The long years of struggle with high-power "spark" were over; now to reap the reward. Marconi fully expected to get the Alternator. When, in 1915, that machine had first been perfected, it had been to the associated Marconi companies that General Electric had turned as a potential market. The Marconi companies had, of course, recognized the value of the machine immediately. Marconi himself, and Mr. Sidney St. J. Stedman, the British Company's counsel, had come over to the United States in that year to negotiate an agreement with the Schenectady electrical firm; they had been well on the way to a final understanding when Italy had called her son to the colors and Marconi had had to leave. The matter had been postponed, then, until after the war, and both parties had made a gentlemen's agreement that neither would do anything until conditions were restored to normal, that would interfere with the execution of a contract.

In 1917 this agreement had been cancelled by mutual consent, and General Electric had negotiated directly with the American Marconi Company for the installation of a fifty-kilowatt machine at the New Brunswick station; it had been put in at General Electric's expense for demonstration purposes. The two-hundred-kilowatt machine that was now there, and that the Navy was still using to handle the traffic to the Peace Conference, had likewise been installed at General Electric's expense—they had remodeled the entire station and the antenna. There was no question about the Alternator's efficiency; it had done yeoman service.

Now, in March, 1919, the representatives of British Marconi came over to the United States again, this time to seal an agreement that would give them the use of this machine that had made worldwide wireless feasible. They were not backward—British Marconi was out to dominate the commercial radio communications of the world—they wanted the exclusive rights to the Alternator; they wanted twenty-four of them immediately, fourteen for the American company and ten for the British, at $127,000 apiece . . . and there were to be others, later. . . .

General Electric was reluctant to grant exclusive rights,—thought itself entitled to a royalty on the Alternator's earnings; the negotiations continued for a fortnight or more, both parties gradually coming nearer and nearer to an agreement. News traveled to Washington. It reached the Naval Communications Office. . . .

The Director of Naval Communications, at that time, was the same William H. G. Bullard, now a Rear Admiral, who had first held the post in 1914. Admiral Bullard, just back from France, was a fighter, a patriot, and a man perhaps better versed in communications, in a broad sense, than anyone else in the Navy. He had been working with radio for years—the Naval Communications Service was in a large measure his creation—and he had come to a deep and passionate belief in the value of American-owned transoceanic radio facilities for American use, not only from a standpoint of defense, but likewise from one of commerce. The reasons for his belief were many, derived for the most part from practical experience, from the episodes of those years that led up to the war.

He was not a jingo, although he had been an ardent advocate of government monopoly and Navy operation of radio. But that had been because he had realized, in pre-war days, that without subsidy the system that he believed in could not be brought to pass. Now, however, sensing the futility of that hope, he was prepared to abandon it, and therein lay his strength, for most of his compatriots still clung to it devotedly. It was the end that Admiral Bullard was fighting for—not the means.

Admiral Bullard was in charge of the operation of Navy radio; another man fortunately able to change his beliefs and think forward was Commander S. C. Hooper, head of the Radio Division of the Bureau of Steam Engineering, and in charge of radio *materiel.* These two officers decided that it was their duty to the nation to take some step in opposition to the impending Marconi monopoly of American and world radio; Commander Hooper, then, by long-distance telephone, asked the General Electric Company if it would not hold up the Alternator negotiations until he and Admiral Bullard had seen and had a conference with their representatives.

This was toward the end of the first week in April, 1919. The meeting—they all felt that it was going to be important, but none of them really dreamed what a profound effect it was ultimately going to have on the world's affairs—was arranged; it would be held in New York on the following Tuesday, the eighth.

They met—Admiral Bullard and Commander Hooper for the Navy, and for the General Electric Company a group including big men, E. W. Rice, Jr., President of the Company, Owen D. Young, one of the Vice Presidents, and Albert G. Davis, Head of the Patent Department—and Admiral Bullard explained the circumstance that had led him to interfere in this legitimate business transaction. His plea was that the Alternator not be sold, that that fine machine which had brought about the realization of the long hoped-for transoceanic communication and which was so entirely an American development, not be handed over to the British to be used competitively without our having it to fight back with. He spoke of experiences in the Orient, where British cables were the only means of telegraphic communication and where at times our messages had been deferred . . . it was not a question of right or wrong, but of the plain, unvarnished inevitable happenings of world competition. He spoke sincerely, earnestly, even passionately, of the subject near his heart. He spoke of the closeness existing between British Marconi and the British government.

And his words impressed his hearers.

But they were business men; apart from their own feelings they had a duty to their stockholders. What, one of them inquired, would they do, then? As electrical manufacturers they had spent large sums of money in developing the Alternator and their only hope of getting any of that money back was to sell a number of the machines. There was no existing market for them beyond the associated Marconi Companies, and while it might be patriotic to reject that market, it was not altogether business, was it? They sympathized with the Admiral and appreciated the gravity of the situation, but had he any solution to offer that suggested another course than tossing away several million dollars?

The Admiral had. He had himself given up the government ownership idea, and knew that there must be both commercial and naval communications, each covering its own specialized field; he wanted the two to work hand in hand as they should. Now was the time for a genuine American venture into the field of commercial radio communications, a strong venture, one of magnitude. British Marconi and the British government worked together to the benefit of every Briton; it was time for an American company to do the same with the government of the United States.

The General Electric men looked interested.

"I want *you,*" the Admiral went on, "to form that company, and to build up an American corporation which will be big enough and strong enough to stand up on its hind legs and look the British Marconi Company in the face as its equal."

They gasped. They were business people, manufacturing electrical apparatus and selling it—successfully, to be sure, but that was their business; they were not in the communications field, didn't know anything about it, and had no ambitions along those lines. . . . They sat there, thinking it over. The idea was certainly a solution of the Alternator difficulty; they knew now that if they sold to the British and to the American Marconi Company there would be nothing but trouble along the way. . . .

Mr. Young took the plunge. "All right," he said. "We'll do it." Mr. Rice agreed. Mr. Davis, the lawyer, was deputed to go back to Washington with the naval men and work out detailed plans for the formation of the new enterprise . . . and the Marconi Companies were notified that negotiations for the Alternator were off until further notice. . . .

The significant feature of this interview was not that it brought about any recognition of the possibly deleterious effect upon the United States of continued dependence upon foreign-owned facilities for international communications; nor that it caused acceptance of the dictum that the radio had now reached full maturity from a standpoint of transoceanic usefulness—it was merely this: that nothing could have been a greater indication of the change in the United States world status than this willingness of her Big Business to enter the communications field. She had not needed cables in the years when cables were being laid down; it had been others that set up those facilities. Now she did *need* world communications. The pioneer nation had merged into world duties and world place. Henceforward she would never be isolated again. . . .

It soon became evident to the gentlemen of the General Electric Company that in the formation of the new radio enterprise they would have to move on their own initiative. They were used to getting things done; used, once they had embarked on a project, to carrying it through to a successful finish with all the resources of their immensely resourceful organization. And it was apparent that Washington was not going to work that way, for though the officers in the Navy Department swung enthusiastically to Admiral Bullard's plan and the Assistant Secretary of the Navy, Franklin D. Roosevelt, was strongly for it, the Secretary, Josephus Daniels—now abroad at the Peace Conference—held matters up.

The initial plan had been for the drawing up of a contract between the Navy Department and the General Electric Company which would define the semi-official status of the new undertaking; this, of course, had called for the sanction of Secretary Daniels. When, on May 25th, after his return from abroad, that official was approached, he was found still to be irrevocably committed to the forlorn hope —government ownership. He was taking the matter to Congress again. . . . He doubted, furthermore, that he could exercise a war power—we were still technically at war with Germany—to sign a contract which was intended to carry on into times of peace. In short, politician though he was, he quite failed to see the political signs of the times, and the utmost to which he would commit himself was the statement that if the Congress rejected government ownership but did give him authority to deal in the matter, the contract would be acceptable. . . .

This, though it was better than nothing, meant, of course, months of waiting—and now, if ever, was the time to strike. Backed by the Admiral, the Company decided to go ahead on its own initiative, and so notified Secretary Daniels. All the features—such as a permanent guarantee that control of the new undertaking would rest in American hands, and a provision that a representative of the Government should sit on the Board of Directors—that were to have been included in the contract with the Navy, would be made a part of the by-laws; the result would be practically the same.

The program was simplicity itself. They proposed to buy the American Marconi Company, lock, stock and barrel! Thus, at one stroke they would acquire patents, stations, an organization, would eliminate the Briton, and take care of 23,000-odd American stockholders in the Marconi Company. The one possible obstacle was the still to be determined attitude of the British stockholders, who controlled matters, and it might be a tough nut to crack. As far as the personnel—from the President down—of the American Marconi Company was concerned, they rejoiced at the prospect. For so many years they had been subject to opposition in the United States that the idea of having a new national character firmly established once and for all, seemed like smooth sailing after troubled seas.

In June, E. J. Nally, American Marconi's Vice President and General Manager, sailed for England with Mr. Davis of the General Electric Company, to take the matter to the Marconi stronghold . . . and during two months of business fencing, of diplomacy and tact they carried on the negotiations side by side. And in the end succeeded! When they returned in September, Mr. Davis brought with him 364,826 shares of American Marconi stock, purchased from its British owners by the General Electric Company; millions of dollars had been paid for it, directly or indirectly, and now, at last, control of American radio lay entirely in American hands.

Step by step the plans went ahead. It must be remembered that there was not a communications man in the General Electric organization; during those days their officials were intently concentrating on the subject—weighing the courses suggested by Admiral Bullard, working, too, in very close harmony with Mr. Nally and the other men of the Marconi Company. When the proposed articles of incorporation, and the contracts to be signed with American and British Marconi were ready, the General Electric Company once more asked Admiral Bullard for assurance that they would have the complete government and Navy backing that he had promised them for the new venture; without it, even now, they would not feel like going ahead. He gave it to them, unreservedly.

Very well, then; the die was cast. On the seventeenth of October, 1919, the new company was organized under the laws of the state of Delaware. It was called the Radio Corporation of America, and it had a capital of five million shares of preferred stock at a par value of $5.00 a share, and five million (shortly increased to 7,500,000) shares of common stock without par value—it was not a small undertaking! It was formed, said the articles of incorporation: "to send and receive signals, messages and communications; to create, install and operate a system of communication which may be international; to improve and prosecute the art and business of electric communications; to radiate, receive and utilize electro-magnetic waves; . . ." etc., etc.

Admiral Bullard's suggestions for a guarantee of American control found fruit in its fundamental laws as follows: "No person shall be eligible for election as a director or officer of the corporation who is not at the time of such election a citizen of the United States. The corporation may by contract or otherwise permit such participation in the administration of its affairs by the Government of the United States as the board of directors deem advisable."

And to carry out those laws, the by-laws provided that not more than twenty per cent of the stock might be held and voted by foreigners without restriction; that those shares were to be marked on their face "Foreign Share Certificates,"—all others were to be transferable only to loyal citizens of the United States or corporations free from foreign control . . . and the Secretary of the Navy or his agent was to be privileged to challenge the vote of any such stock if the contrary was believed to be the case. The board of directors invited President Wilson, then, to appoint a government representative to sit with them at their meetings and participate in the company's administration . . . and the President appointed Admiral Bullard. . . .

There. The blade was forged; now to make use

of it. Somewhat to the disgruntlement of the younger element in the Navy Department—such was the old sentiment against the Marconi Company—E. J. Nally, who as Vice President had been steering American Marconi through these last troublous years, was made first President of the Radio Corporation,—an excellent choice, because of his administrative ability, his knowledge of conditions in England and Europe, and his long contact with the directing heads of British Marconi. He would be invaluable.

And on the twenty-second of October a preliminary agreement was signed by the General Electric Company, the Radio Corporation of America and the Marconi Wireless Telegraph Company of America; the latter agreed to attempt to persuade its stockholders to consent to the company's sale under the terms stipulated. . . .

On November 20th, almost twenty years to a day after young Guglielmo Marconi and his business associates had caused it to be formed, the Marconi Wireless Telegraph Company of America sold its assets and its stations and its patents, its contracts, its good-will and its going business, to its lusty new successor; two million shares of RCA common and two million of RCA preferred were involved in the transaction—they went to the Marconi shareholders. . . . There were certain claims held by the Marconi Company against various parties—the United States Government, for one—for infringement of patents; these were not parted with, and so that they might be prosecuted the Company did not dissolve. But apart from that it was out of business. . . .

On the same busy afternoon that saw this momentous deal take place, the new Radio Corporation and its parent, the General Electric Company, had an understanding in the form of a General Agreement to govern their future relationships. They exchanged the right to use each others' patents and those which they might acquire; the General Electric Company agreed to make and sell radio apparatus covered by these patents exclusively to the Radio Corporation, and the Radio Corporation agreed to buy radio apparatus covered by these patents exclusively from General Electric. . . .

It had the effect of making the new Radio Corporation the owner of virtually all the commercial high-power radio facilities in the country (though it could not operate them as yet, for they were still in the Navy's hands), and it made it the General Electric Company's American sales agent for radio apparatus. The parent was to remain a manufacturer; the subsidiary was to be this particular company's first venture into the operative field.

Between them the two companies had: (1) all the Marconi patents, including the important tuning patents and those of the Fleming Valve that covered, fundamentally, the use of vacuum tubes; (2) the Alexanderson Alternator patents; (3) the

Alexanderson multiple tuned antenna patents; (4) the Langmuir "pure electron discharge tube" invention, patent applied for but not yet granted because of the interference with Arnold of Western Electric; (5) the basic patents on tungsten filaments; and (6) those of Coolidge covering drawn tungsten wire. . . .

American Big Business, now, was in radio. Apart from this active and wholesale entry by General Electric into the field, there was Western Electric —owner of the De Forest Audion and other patents —preparing to make active use of the art for the Telephone Company's purposes; an experimental circuit was going in between Catalina Island, off the Southern California coast, and the mainland, whereby the island's telephones would be connected directly with the shore telephone exchanges by a radio link.

And down in Pittsburgh, at the Westinghouse Electric and Manufacturing Company, they were beginning to give serious thought to the situation. Westinghouse was General Electric's great competitor, the second of America's electrical giants. Like the Schenectady firm, it had confined its pre-war radio activity to its laboratories, but what work it had done in them was far less important than the General Electric vacuum tube and Alternator contributions. Nevertheless, during the war it had gotten into the art because it had had lamp-making machinery adaptable to vacuum tubes; and because of government need and the guaranteed patent situation.

Now Westinghouse saw itself very much out in the cold, just as it was beginning to look as though radio might be a pretty good field to be in—if General Electric were going to make so much out of it, the Pittsburghers wanted to join the parade.

There were only two important sets of radio patents wandering about without strong owners. The Edwin H. Armstrong "feed-back" circuit patent, which covered the use of vacuum tubes as generators of continuous waves, and various other important Armstrong and Pupin radio patents, were still the property of Mr. Armstrong and his friends; they were suddenly very important because of the growing use of vacuum tubes, though so far the only business use made of them had been the licensing by Armstrong of a number of small manufacturers of radio apparatus for amateurs.

And the Fessenden patents. . . . It will be remembered that when the war inflation of the radio industry began, the old National Electric Signalling Company, which owned them, had been reconstituted under the name of the International Radio Telegraph Company; since then it had done about two million dollars' worth of manufacturing for the Government, but the war was over and the market was dead. Its greatest assets were these Fessenden patents and some that it had developed by itself since 1912; the Fessenden patents completely covered the "heterodyne" principle. . . .

Two or three months after the formation of the Radio Corporation, the Westinghouse Company approached the International Radio Telegraph Company with a request to be granted licenses under the I. R. T. patents. This request was rejected. The men who had nursed those patents along for ten years, now, were unwilling to see them exploited by others; it had always been their dream to form a communications company. . . . And thus were entered into negotiations whereby these two organizations went into partnership to form such a company, which was to stand in the same relationship to Westinghouse as the Radio Corporation did to General Electric. It was to be known as The International Radio Telegraph Company (the difference in name being merely the inclusion of the "The" as part of the title); the I. R. T. people were to furnish their physical assets and patents; Westinghouse to contribute $2,500,000 in cash as the new company needed it.

The agreement to do this was signed by the two companies on the twenty-second of May, 1920; in June the new company was actually formed, and cross-license and restrictive manufacture and sale agreements put through that bound it to its parent. . . .

But now that they had it, they found that they had not so much. They wanted to go into transoceanic communication on the same scale as the Radio Corporation, but they had no high-power transmitter to do it with. . . . The nearest approach to such a thing was the old Fessenden 500 cycle "spark" idea that had originally been used at Arlington, but in the light of the Alexanderson Alternator that was obsolete. The only alternative was to investigate the possibility of using the "arc," the patent to which was still at the time owned by the Navy. . . .

Down in the Navy Department they found encouragement. The Navy didn't mind how much big business entered the communications field; the more the merrier! And in August granted them a license to use all radio patents owned by the Navy . . . they had the "arc," then.

But they found themselves up against the old fundamental of communications; they had no "station at the other end." The day after the formation of the Radio Corporation that enterprise had signed exclusive agreements to exchange patents and traffic with British Marconi and was reaching out to cover the European field as tightly as possible. International was just six months too late. When its president, Samuel Kintner, went abroad in that summer of 1920 to get traffic agreements with the various European countries, he found the situation beyond remedy; the best he could do was to get a promise in Germany that if he erected a suitable station in the United States he would receive the same consideration as the Radio Corporation. . . .

There was not enough to go on, yet. No, they must seek to strengthen their position. The first step must be the acquisition of more radio patents; patents were weapons in this game, and without weapons there could be no strategy. On October 5th, they bought an option on the Armstrong-Pupin patents; a month later, on the fourth of November, they exercised it, and came into possession of the "feed-back" and others for a sum between $335,000 and $535,000 which was to depend upon the outcome of the litigation then pending regarding the "feed-back." This was the most strategic group of patents they could buy, for coupled with those of Fessenden, it dominated all receiving apparatus using vacuum tubes . . . and the "feed-back" covered the tube transmitter. . . .

Meantime, in those post-war days, the Naval Communications Service had been having its troubles. While it had been lending the support of its good right arm to the formation of a strong American commercial radio industry, it had continued, as during the war, to carry on the radio communications of the country, and had found the job a very different one now that peace had come.

It was beginning to see that whereas during the war the cream of the country had been willing, even anxious, to jump in and lend a hand without regard to returns, the very men it had come to depend on were now getting out of the service as fast as that could be done. There was this enormous number of stations, all of which must be operated to provide proper service, and which required several thousand "radiomen" to do nothing more than stand watches; as these men had been needed, they had been trained and provided, but now . . .

In June and July, 1919, orders went out from Washington that all men who had enlisted subsequent to the declaration of war on April 6th, 1917, were entitled to be discharged. Politicians, who had been appealed to to "get the boys out of uniform," were clamoring. In one dizzy swoop the wartime naval force of over half a million trained men slid relentlessly down until only a hundred thousand odd were left, and many of these were new recruits! Away went the radio operators, and there was nothing that could be done to hold them. Would they stay and work for forty or fifty dollars a month? They would not, not with civilian life and freedom calling.

And the officers who had built up that service—which, for all its twenty-five million dollars' worth of apparatus, was only as good as its personnel—watched it crumble, watched it change from a splendidly efficient body to one which was frankly unable to handle its job. The good men were concentrated, as much as possible, at the most important stations, but the complaints began to come in in increasing numbers. Such and such a vessel had called a station for hours! What was the matter with the Navy operators? Asleep on the job? No. There hadn't been any operator. No men.

Under the circumstances, they performed prodigies. They laid the foundations for the new, peacetime Naval Communications Service and got to work training new men; they carried on with the ship-shore work. The commercial overseas work was turned back to commercial facilities—the cables—but still the Navy operated the East Coast high-power radio stations and those of its Pacific chain, and the millions of words of traffic between Washington and France incident to President Wilson's trips abroad and to the Peace Conference were all handled, and well. On the Pacific, due to the ban against competing with commercial facilities, the high-power chain was closed to commercial traffic and the Mackay cables were left to handle it . . . until the press of messages was so great, in the autumn of 1919, that the cables were swamped, a week, ten days behind in messages. In Honolulu they were complaining that it was quicker to send letters. . . . Congress directed, then, that the Navy reopen its stations to commercial messages.

And ordered the Navy to return all the privately owned high-power radio stations to their owners as soon as possible. There were only nine, all now the property of the Radio Corporation of America.

Early in 1920 they were returned—the two in

Massachusetts, three (including Tuckerton) in New Jersey, two in California, and two in Hawaii. . . .

And with their passage from the Navy's hands, business again took over business, put its shoulders to the wheel and got to work. On March 1st the Radio Corporation of America transmitted its first commercial messages to England, from New Brunswick. The cable rate had been 25 cents a word; these radio messages cost only 17. On the same day it opened service to and from the Marconi station at Stavanger, Norway—cable 35 cents, radio 24 cents a word. In August it opened to Germany; in December to France. . . .

Normalcy. . . .

PART THREE

THE ERA OF POPULAR USE

CHAPTER NINE

BROADCASTING EMERGES

In those years 1919 and 1920 people were trying to understand what had happened, what was happening to America. Tranquillity was gone. The old placidity of life had been first disturbed by the bloodshed in Europe (the news-reels had made that seem, somehow, akin to the fantastic cinema world that accompanied its portrayal on the screen), and then snatched out of itself by twenty months of war effort with a background of sacrifice and spiritual exaltation.

That exaltation was evaporating, now. The war motive that had lent purpose to every unreal action no longer existed—but it was impossible to turn back and resume precisely the pathway of life that had gone before. Too many events had intervened; too many new habits been formed. A new and complicated environment had all too perceptibly grown up in America—and instead of being able to put it aside, America found herself being adapted to it.

Perhaps the most striking interior change in the nation had been the growth of intercourse between the people. Drawn together by the unifying effect of a common effort, the prosecution of the struggle had shaken man out of his own community. A sectionalized America had been made to get up and out, to travel. The use of every form of intercommunication had spread. The demand for news had stimulated the news-distributing agencies, both film and press, to unprecedented activity. The demand for personal movement had stimulated the use of the automobile, hitherto something of a luxury, and made it a commonplace. The exigency of the moment had made money loose—rather than tight, as in other countries—as if there were boundless confidence in the ampleness of the nation's resources. Hoarding had been almost unknown; it had been a lavish era. . . .

And so the men who returned from overseas found a different nation. The change was most visible in the cities, in the restlessness of people, in their incessant movement—streets that had been dignified, sedate, were jammed with motor traffic, the fumes of gasoline were in the air, and there was a great unease . . . people who had learned that they could work as they had never worked before were many of them reaching out to play with the same vigor—life was taking on a new pace. Sport was spreading incredibly, both for amusement and for exercise. Woman was participating in the world's affairs and had dropped her Victorian bonds. All the things that had hitherto been the province of the wealthy were spreading through other strata of life—country clubs, motor cars, golf, tennis, wrist watches . . . ironmongers were wearing silk shirts; there was a dance mania and a jazz mania . . . and a nation that had so much vitality that it didn't know what to do with it was bubbling over at every crack and cranny. . . .

Nobody understood. There was a curious sentiment that, now that the spiritual exaltation of the war had evaporated, everything must be wrong. People, groping to comprehend the era, felt that youth had got completely out of hand, that everything new was a sign of degeneracy, and that the only hope for mankind was a return to puritanism . . . but instead of going to church they went automobile riding; the wartime edict on alcoholic prohibition, born of the spiritual exaltation, was simultaneously lauded and violated; bigots censured and tried to censor everything, sometimes succeeding, sometimes aggravating the very things they decried.

The most striking environmental feature of the era was the availability and practicability of instruments with which man could combat nature personally—particularly nature's physical barriers. Of these the automobile was, of course, the most prominent. In 1914 it had still required something of a skilled mechanic to embark upon an extended voyage in a motor car; in 1919 it "got you there and it brought you back," whoever you were—that summer and the next thousands of Eastern people started for California, and thousands of Western people started for New York, because they liked to motor and wanted to see what was on the other side of things; a gypsy-like phenomenon called "auto-camping" made itself evident and proceeded to grow by leaps and bounds; the old migratory instinct of the American people that had induced the original westward movement, came forth again —it was a strong strain of independence and a strong faith in their ability to "find a job" and make a living wherever they went . . . cities like Los Angeles doubled their population in half a decade because people "liked the climate" and had this amazing new freedom to move themselves whither they listed. . . .

Before the war there had been some five thousand licensed radio amateurs scattered throughout the nation, most of them youngsters who had tinkered with "spark" radio. Limitations on power and wave length had made the achievement of great distances generally impossible by them, but through their organization, The American Radio Relay

League, they had been able to exchange communications from coast to coast. And they had filled one most important place in radio activities,—they had served as a coöperating audience to the serious experimenters who were striving to perfect the more subtle utilizations of the art—continuous wave generation by vacuum tubes, and radiotelephony. The amateurs, with their reports, had furnished invaluable information as to the quality and strength of the signals, so that men like Armstrong and De Forest and the workers in laboratories could tell what results they were getting. . . . The war, however, and the ban on privately owned stations, had ended all that.

In mid-1919 this ban was removed. Now, added to the former amateurs, were the thousands who had been trained in radio communications at Harvard and on Navy ships; there were fifteen or twenty thousand men in the country with a working knowledge of the art. Within a year six thousand of them had taken out amateur licenses again and had their stations going; telegraphy was still their chief interest—they had transmitting as well as receiving equipment. . . .

But the radio art was new. Its chief characteristic before the war had been that it was a "fiddley" proposition, a thing of adjustments and tinkerings, of finding a good "spot" on a crystal detector, of intricate "tuning" by means of long coils that slid one within the other. During the war this had changed. The new vacuum tube and its new circuit arrangements were precise, exact, always identic in performance. And now all the amateurs wanted to use the new radio instead of the old, but there were no vacuum tubes to be had . . . except as, for example, they were pilfered from government supplies and sold for fancy prices . . . and though there was a desire to work along the new lines of the art, it was irritatingly checkmated. . . .

The very difficulty was an incentive. . . .

The amateur was an enthusiast. Radio was an adventure to him. He knew what it was to put a collection of inanimate wires and coils and condensers together, to string up an antenna, and then to hear that mysterious *peep-peep-peeeeeeeeeeep* in his telephones, plucking it right out of nothing, able to translate it into the thought launched into space by another person sitting in an unknown place . . . it was an unending thrill. A thrill to reply, too, to that unknown communicant. And when it was a voice! Before the war the radio voice had come from out a roar—the steady crash of an "arc," and in the middle of it that quiet little procession of words . . . when it was a voice and the amateur had friends about, there would be an excited passing of head-phones from one to another. . . . "Listen! You can hear a man talking!"

There was, in one American city, a new voice in the ether in those days when 1919 was passing into 1920. Just a man experimenting, as hundreds of others had experimented, but there was a difference

. . . this voice came clear and strong. There was no roar. It did not sound experimental, but fully satisfying—the telephonic branch of the art was over the crest! Like the automobile, it had "arrived"—it was practicable. . . . The man behind that voice was unwittingly serving as the trigger that let loose an avalanche.

His name was Frank Conrad. He was an engineer in the laboratories of the Westinghouse Electric and Manufacturing Company at East Pittsburgh, Pennsylvania, and he was working on the problem of developing a special radio telephone transmitter for the Navy. He had set up an experimental station in his home, using Navy vacuum tubes—because the Westinghouse Company was unable either to make or buy them, due to the frozen patent situation—and as he made improvements and discoveries in his work, he tested them there by playing victrola records. There was a radio audience, at first, of fifteen or twenty amateurs with home-made sets. They heard him once, twice—they began to listen for the Conrad tests. There was a peculiar thrill in hearing the electric word, sung or spoken; it was irresistible. One had to have a particular kind of temperament to be drawn to the telegraphic signal, but everyone was attracted to the word. Those amateurs' friends, having heard it, got an itch to participate. The thought that that thing was coming through the ether, pleasingly adequate, and that they, who might be hearing it, were not. . . .

They wanted to know how to hear it. Some of

them got circuit diagrams, bought parts, and built crude receiving sets for themselves; some of them got their amateur acquaintances to build them. These people didn't care about sending . . . they wanted to hear.

And then a most significant thing happened to Dr. Frank Conrad. The pushing, sturdy, sometimes vulgar, always strongly human voice of the new American public came right into his workshop in answer to the electric word he had sent out of it, and he found his tests taken irresistibly out of his hands.

For he began to get letters. With thanks, occasionally with criticism, often with suggestions. Would he do this? And that? Didn't he think it would be a good idea to try so and so? Why couldn't he play his radio music *every* night? Why couldn't he announce what a record was going to be before he played it? Why couldn't he get some new records? . . . Some of his listeners went further, even, than that—they bundled up their own records and sent them to him! And in the end, in response to this popular pressure, the tests that this individual engineer was carrying out as part of his employers' work for the government of the United States were held, not at his pleasure, but at that of his auditors—which had a certain element of justice in it, since without auditors, the tests meant nothing.

On Wednesday and Saturday nights, then, he broadcast his little programs. Inevitably there grew up between him and that invisible audience a certain bond, a personal acquaintance. They, who had never seen him, came to know his voice, something of his personality. And he who had never seen them, got their letters and remembered, as he changed his phonograph records, that they were out there somewhere . . . he wondered whether or not they would like this one. . . .

It got into the press.

By the late summer of 1920 the Pittsburgh Department Stores had begun to advertise "approved radio receiving sets for listening to Dr. Conrad's concerts." And at that the Vice President of the Westinghouse Company, Harry P. Davis, who had been watching the playing of the whole little comedy, pricked up his ears. Manufacture and market go hand in hand; if a market for receiving sets could be stimulated by this broadcasting . . . and if the very fact that it was being carried on induced publicity and good-will for the broadcaster . . . mightn't it be a good thing for the Company to think of going into? A conception took shape in his mind. Why not make this more than victrola records? Why not make it a "Newspaper of the Air," or something like that . . . with a musical accompaniment? Phrases began to form . . . "the voice of Pittsburgh" . . . "the town crier" . . . it captured his imagination.

Suppose they tried it; it would be well to launch the innovation as spectacularly as possible. The elections of 1920 were coming on; Harding and Cox were campaigning. He thought of the crowds that would gather around the newspaper offices to watch the election returns thrown on a screen—why not extend the returns to the radio audience? And after that give nightly programs, announcing them beforehand. . . .

His mind was made up. This was in the middle of October. He told Dr. Conrad of the decision, and a new telephone transmitter was hurriedly constructed in a building at the East Pittsburgh works; they had scarcely an opportunity to test it before Election Night, and on that night, November 2nd, the engineer did not dare stay, but went home to his own transmitter and "stood by" ready to take up the broadcasting in case anything went wrong—the program *must* be got through.

Nothing went wrong. There was a telephone line from the *Pittsburgh Post,* and as the returns came in they were put on the air. The radio audience numbered perhaps between five hundred and a thousand. . . .

At the time, the station was licensed merely for experiments. Later, when a permanent transmitting apparatus was installed, it was assigned the radio call letters KDKA by the Department of Commerce, just as though it were a ship station. . . .

Meantime the radio industry, which had seemed to get off to such a promising and well launched start after the war, was having all sorts of difficulties. The trouble was that the new radio, which had been developed during the war when it was possible to pool all the patents of the art, was now denied to any one company . . . its crucial devices were distributed among the three big manufacturers which had interested themselves; the result was a perfect stalemate.

First and most acute was the vacuum tube difficulty. The pure electron discharge vacuum tube was the device about which the whole of the new art revolved; everyone that had anything to do with radio wanted it . . . and no one could get it. It was the device that had made radiotelephony commercially possible; as a detector and amplifier it had revolutionized the reception of Hertz waves; it was beginning to do the same for their generation . . . and it was already widely in use in the wire-telephone lines of the country.

The deadlock lay in that the General Electric Company and the Radio Corporation of America owned the Fleming two-element Valve patent, the Langmuir "pure electron discharge tube" patent application, and a number of patents and applications on the interior construction and use of the new tube as developed in the General Electric laboratories . . . while the American Telephone and Telegraph Company and Western Electric Company owned the De Forest three-element Audion patent, the Arnold "high vacuum tube" application (in interference with Langmuir's application), and likewise a number of patents and applications on interior construction and use. . . .

This deadlock had not hampered the development of the tube, for the laboratories of both groups had gone ahead improving the device and making it for experimental purposes. Western Electric and the Telephone Company had made free use of it in the long-distance telephone lines, knowing that a day of reckoning was coming, but willing to leave that to the lawyers. . . .

It was the public that was suffering, for had any of the companies taken the plunge and placed the device on sale, an instant injunction and suit for infringement would undoubtedly have resulted. Meantime the Navy was running out of its wartime supply of tubes, and was observing that the lack of them was growing to have an effect upon the safety of life at sea—vessels equipped for tube transmission and reception were in danger of having their sets put out of commission.

Everyone knew that the situation was intolerable—none better than the companies concerned, and when the Navy began urging them to get together and compromise the matter, and followed these urgings with a duplicate official letter to each group (on January 5th, 1920), saying: ". . . it is a public necessity that such arrangement be made without further delay, and this letter may be considered as an appeal, for the good of the public, for a remedy to the situation . . ." . . . they accepted this governmental mediation and their representatives met.

It was difficult, in the negotiations that followed,

to draw the line at vacuum tubes. What was recognized was that the four companies ought to work hand in hand, each in the main continuing in its respective field, but under some sort of a flexible coöperative agreement on common sense lines. They consented, then, to grant each other the right to use all each other's radio patents, and on July 1st, 1920, a long legal document to these ends was signed by President Rice of the General Electric Company and President Thayer of the American Telephone and Telegraph Company. Its provisions were extended to the Radio Corporation and Western Electric on the same day. In order that each group might protect its legitimate business and to prevent misunderstanding, this agreement stated that as far as their communications activities were concerned, the Telephone Companies were licensed under these patents only for radiotelephony and General Electric for radiotelegraphy; it assigned the manufacture and sale of radiotelephone broadcast transmitters under these patents as within the province of the Telephone Companies, and the manufacture and sale of radio receiving apparatus under these patents for amateur and general purposes as within that of the General Electric Company and Radio Corporation. Each party had the right to make vacuum tubes for use in its own work, and most important of all, each of them agreed to a sweeping exchange of radio information, and to the future coöperation of their research laboratories.

Now, for the first time in the history of the radio art in the United States, a three-element vacuum tube could be legitimately made and sold for general use! The De Forest Audion had been patented in 1907—ever since that date there had been three-element tubes of some sort known, but barred to the public except as illegally manufactured and "bootlegged." Here at last, after all sorts of trials and tribulations and an enormous amount of court action, the tube deadlock was broken. The General Electric Company immediately commenced the manufacture of the device for amateur use under the trade name *radiotron,* and within a short time they were on sale. . . .

In August the American Telephone and Telegraph Company bought from the Radio Corporation of America 500,000 shares of common and 500,000 shares of preferred stock; two of the Radio Corporation directors, then, represented the Telephone Company. . . .

The laboratories of the two groups, coöperating, not only made constant improvements in detector and amplifier tubes from that time on, but began that progress in the creation of "power" tubes for use as generators of continuous waves that within five years was to result in a water-cooled affair that would handle 100 kilowatts—a pair of them equal to an Alexanderson Alternator!

But Big Business, as it put its laboratories to work

and carried on activities in its new radio-communications ventures, found itself still far from free. The difficulty, now, was the schism caused by the separation of the vacuum tube's ownership from that of the modern circuits involving its use. As well have a pot full of raw meat in one room, a lighted gas range in another, and a legal barrier between the two . . . coupled with an enormous appetite.

The Fessenden and Armstrong-Pupin patents, this time, were the sore spot. The Westinghouse Company had indeed chosen wisely when it selected these as its wedge for entering into activities in radio—for they covered all receiving apparatus using vacuum tubes. The Radio Corporation, operating radio-communication facilities that demanded the last word in the radio apparatus of the day, was forced to use receiving circuits that were literally obsolete, or face infringements suits. Early in 1920 they approached Westinghouse and The International Radio Telegraph Company, to inquire the possibility of securing licenses under the Fessenden and Armstrong-Pupin patents.

By this time the International Company had made its disappointing attempt to enter the transoceanic radio field; balked in this attempt the Westinghouse people were working ahead on the active development of their new broadcasting activity—and incidentally doing what the Telephone Company had done earlier, infringing patents right and left regardless, knowing that a day of reckoning would come but not caring because it was development and experimental work they were doing—and in that connection had worked out four sets of complete receiving apparatus, all involving tubes, which they wanted to sell to amateurs. Never was the hopeless state of the patent situation better demonstrated. The circuits of these sets, although they were fundamentally the Armstrong regenerative circuits, would not operate properly without the use of certain minor "hook-ups," the patents to which were owned by General Electric and the Telephone Company; the circuits likewise depended, of course, on the use of tubes.

As the experience with the Telephone Company had shown, there was one way of cutting the tangle at a stroke. On June 30th, 1921, then, the same sort of cross-license agreement that had ended the tube difficulties, was entered into between Westinghouse, General Electric and the Radio Corporation, and its provisions were extended to include the Telephone Company and Western Electric. Incident to this transaction, The International Radio Telegraph Company, whose chief assets were its patents and whose activities had never amounted to much, was sold to the Radio Corporation of America, so that from then onward the Radio Corporation filled identic functions for both Westinghouse and General Electric. It was their radio operations company; it was their radio sales agency. They themselves remained merely manufacturers. And since they had not contributed equally to the Radio Corporation's development, they mutually agreed that sixty per cent of its orders for radio material manufactured under the cross-licensed patents were to go to General Electric, forty per cent to Westinghouse. They pledged themselves not to manufacture radio apparatus covered by patents under which the parties had cross-licensed each other except for the Radio Corporation and for the Government (to whom they might sell direct); the Radio Corporation, in turn, was pledged not to buy radio apparatus covered by these patents except from the parent companies—it acquired rights to use and sell this apparatus.

Payment for The International Radio Telegraph Company was made in Radio Corporation stock, and thereafter Westinghouse was represented on the board of directors. In 1922 it held 1,000,000 shares of common and 1,000,000 shares of preferred stock in the Radio Corporation of America.

At the time, the Armstrong-Pupin patents were not entirely paid for; the Radio Corporation and General Electric, which were of course to have the use of them, agreed to share in the purchase price.

Thus the air, and a patent situation probably more complicated than any in the previous annals of industry and involving some 1200 patents, were cleared completely, for just a short time previously, on March 7th, 1921, the United Fruit Company and its Wireless Specialty Apparatus Company had entered the fold. The Wireless Specialty people had had, apart from the crystal detector patents, one that had a bearing on the Fleming Valve—it had been involved in a suit against the Marconi Company which still hung over and was pending against the Radio Corporation. United Fruit, likewise, through the Tropical Radio Telegraph Company, had a complete system of radio stations (spark equipped) for communication to Central America, and a Marconi Company contract to provide certain apparatus . . . the subject of much dispute, and now carried over to the Radio Corporation. In order to end the litigation and to make possible the modernization of Caribbean radio without duplication of stations and elimination of United Fruit's vital service (still a money loser), the General Electric Company bought some fifty per cent of the stock of the Wireless Specialty Apparatus Company, and cross-licenses were exchanged between that company, the Radio Corporation and the various manufacturers interested in the Radio Corporation. Payment for the Wireless Specialty stock was made in cash; shortly thereafter United Fruit bought 200,000 shares of common and 200,000 shares of preferred stock in the Radio Corporation, and was represented on the board of directors. . . .

By the end of 1922 the Radio Corporation of America was truly America's expression in radio, and a forty million dollar concern; behind it were the General Electric Company, the Westinghouse Electric and Manufacturing Company, the American Telephone and Telegraph Company, the Western Electric Company, the United Fruit Company, and the Wireless Specialty Apparatus Company—a total of several billion dollars in resources, the most advanced thought in business and scientific research in the nation, and some of the nation's biggest men. All of these companies had contributed patents which, used alone, could neither do them good nor make money for them; pooled together, they created an art. The Radio Corporation was their link, and would serve as the instrument through which each of them would get a proportionate return for what he had contributed. . . .

The way was prepared for the astonishing thing that was to come and which men still did not fully realize was to happen.

While American Big Business was thus coöperating to free radio from its deadlock of patents and cross-purposes, several interesting things had been taking place—predictive of the era to come. In May, 1919, the United States Navy "NC" seaplanes, and in July the British dirigible R-34, had made extensive use of the radio as an adjunct to aerial navigation in connection with their pioneering flights across the Atlantic . . . all the progressive nations of the earth were enacting laws similar to those of the United States requiring seagoing ships to be radio equipped. . . . General Squier of the Army Signal Corps was making discoveries that Hertz waves could be made to follow wires, not travelling "inside" as ordinary currents do, but clinging to their surface—he called his system "wired-wireless," and one of the most important things about it was that it permitted "multiplexing," the sending of more than one message along the guiding wire at one time, hitherto impossible in telephony. . . . John Hays Hammond, a New York engineer, was perfecting new apparatus for the distant control of machinery by radio; using it, the obsolete battleship *Ohio* was made to steam at different speeds on varying courses while other men-o'-war fired their heavy guns at her. . . . New knowledge of radio waves was teaching men that they knew almost nothing about "sonics"—the behavior of sound waves. . . . There was a feeling that startling things were on the verge of happening, and in dozens of laboratories thousands of men went forward on work that now progressed by leaps and bounds.

In Pittsburgh, the radio audience was growing. The Westinghouse Company was making plans to establish two new "newspapers of the air," at Newark (WJZ) and at Springfield, Massachusetts (WBZ). It was strange, how the thing drew listeners. In Washington the Naval Aircraft Radio Laboratory at Anacostia had begun to broadcast experimentally; the response had been so great that they had been carried away by it, had moved in a piano and were getting speakers and singers. The pressure of the radio audience, it seemed, was irresistible. The Navy Department realized with something of a shock that the laboratory was doing more broadcasting than research . . . and ordered them to get back to business. . . .

CHAPTER TEN

THE RADIO BOOM

In the seven years between the dawn of 1921 and the dawn of 1928 the popular use of radio spread as nothing before has ever spread, not only into every nook and cranny of the United States, but in growing waves all over the earth. On the former date there were not over six or seven thousand privately owned sets of radio receiving instruments in this country; at the end of those seven years the number was on the order of ten million, three quarters of which were "tube" sets of a high standard of scientific perfection, and the radio audience of a few thousand had grown to number, on occasion, more than half the adult population of the land. It was possible, then, by interconnection of transmitting stations, for that stupendous concentration of human minds to be focused upon a single voice, a single instrument, a single event. . . .

It is perhaps difficult for those who have lived through this change to comprehend what it signifies in terms of world history. No nation ever had greater communication barriers than those of the wilderness infancy of the United States; no nation has ever so broken those barriers down or achieved such astonishing unanimity and rapidity of thought conveyance as has this one in its young maturity. It took aeons of time for the use of fire to spread among men, aeons of time to develop a substantial man-made structure to shelter him from the elements, aeons of time for him to learn to speak, other aeons to write—his progress along the pathway up from brutehood has been painfully, pitifully slow . . . and now, in this new era of science and intercommunication, of which these United States are such a vital expression, an entire nation has come to the point of absorbing some new thing into its life, a thing that will henceforward play a profounder part in its environment than it can guess, in the short span of a little more than two thousand days. . . .

There were, of course, reasons why the radio should spread. It captured the imagination. It brought more news to a news-avid people. It brought music to a nation with a distinct, though untrained and unsophisticated, hunger for music, and which, since its environment had so far precluded that it should produce musicians in any quantity, had turned widely to the phonograph—one great reason for radio's initial adoption was its inherent superiority to that instrument in that it offered a directer sense of contact with the artist; it eliminated the "record" with its mechanical repetitions that came to destroy any sense of freshness, of illusion. Above all, in quality of reproduction the radio was from the beginning of the vacuum tube era, equal if not superior to the line-telephone and the phonograph, which had set the standard. To man, as he first heard it, it was something new and something better, and though it did not and could never replace the older agencies of music, news or communication, he had been educated to the point of instant recognition of its special features.

The thing grew with amazing rapidity. From the moment it was released by the clearing of the patent situation in July, 1921, it started to appear simultaneously in all parts of the country. That summer and fall the masts and antennae of broadcasting stations were going up from coast to coast. In September, 1921, the first three broadcasters to follow "KDKA" went "on the air." There was another in October, one more in November, and then in December there were twenty-three . . . eight in January, 1922, twenty-four in February, *seventy-seven* in March . . . by the end of the first year there were five hundred and eight stations broadcasting in the United States, and coupled with this mad rush to set the still ether in motion had been an equally phenomenal rush on the part of the public to procure some sort of receiving apparatus . . . the radio audience had leaped to a million. . . .

It was a strange situation, sprung as it had from nothing. Radio had become the fad of the hour. People, particularly the broadcasters, had rushed into it believing anything and everything of it—that it was the road to wealth, fame, happiness and what not—eager to participate from all sorts of motives from money-grabbing to sheer altruism, lacking any knowledge of the fundamentals either of the art or of what later grew to become a highly specialized broadcast technique. Only a handful of those first broadcasters—some thirty or forty of the five hundred—turned to the recognized agencies for their apparatus; the rest, instead of looking to the companies whose laboratories had developed the art, got hold of any Tom, Dick or Harry who had "learned radio during the war," to put their station together. Sometimes the results were unexpectedly good; more often they were horrible—but in any case as soon as a station was "on the air" the audience appeared.

For the moment anything in the way of a program sufficed. Few broadcasters had any idea what they were entering into, or realized that if broadcasting was to endure the laws of business must be applied to it. Many of them believed that their initial investment — invariably underestimated— would be the last; they had no realization what an inexorable taskmaster, what a challenging puzzle, their great invisible audience was going to be once that "fad" moment was over. . . .

An audience is one of the most valuable things in the world, and that fact holds particularly true in a democracy where it has the freedom and power to act. An audience has, likewise, purchasing power. The money which it has to spend can be made to flow more easily, or diverted to particular channels, if it can be reached with convincing argument that will serve that end.

There were already in the United States means of reaching the national audience, both in whole and in part. The public mind was sought by a highly developed and widely circulated press, in the form of daily news mediums and of popular magazines. These mediums were used to some extent as means of swaying public opinion, but much more as carriers of "advertising," which reached for the public pocket-book. Advertising had become one of the largest activities in America, and it had served as a powerful influence for the leveling of partisan tendencies on the part of the press. The large amounts of money which the advertisers had to spend flowed toward the press-mediums that had the greatest "audiences"; a Republican dollar and a Democratic dollar had precisely the same purchasing power, and the press was tending more and more to become a straight purveyor of news and "literature," and less and less an agency that tried to interpret events or attempted to sway public opinion.

There had been a time when this demand for an audience had been destructive of quality. The largest crowd can be collected in the shortest time by the sensational, the thrilling, the shocking. . . . There had been an era of "yellow" journalism which had built up enormous circulation by stressing the bizarre and sordid in American life,—its survivors still existed as "tabloids"—but the advertisers had come to shy from those mediums because they trended toward the moron-reader, and that was not the reader who had the money. There had grown, then, strong efforts toward quality circulation, and with them the development of a truly astonishing national service of accurate news gathering and dissemination through the *Associated Press* and the *United Press,* and of "syndicated" feature services of a somewhat higher order than those of yellow journalism. And there had grown several rather remarkable weeklies and monthlies with national circulations, which, though they brought few masterpieces of literature to the masses, at least achieved a reasonable standard of literary merit. It was recognized that to reach and hold large numbers of substantial citizens—which in America included everybody with a regular income—a medium must contain an admixture of facts, entertainment, and "color" . . . the *Saturday Evening Post* had its articles, its stories, its illustrations and humor . . .

the *Daily Gazette* had its local and A.P. news, its "home" page, and its comic strip . . . each of them had advertising in direct proportion to circulation. . . .

It was advertising that had come to have a good deal of subtlety. Some of it frankly presented an appeal to buy "the best"; other merely served to keep a company's name before the public.

If it had not been that audiences were so valuable, the radio "fad" might have died a natural death after that first unreasoning year. Certainly that would have been the result if it had been wholly dependent upon those first five hundred broadcasters, for two hundred and ninety-five of them gave up the ghost in the next twelvemonth, having bitten off more than they realized and suffering from acute contraction of the purse.

But the audience was not to be allowed to abandon its receivers. From the moment it was realized that there existed a million and more American people who were willing to sit and listen to the sounds that came out of radio receivers, there appeared those willing to put time, brains, and money into the dual problem of keeping them there and inducing others to join them. During that second year when two hundred and ninety-five broadcasters dropped out, three hundred and fifty new ones sprang up to take their places . . . and the audience, playing a most willing part (because those intent on capturing it were using entertainment as the offering) doubled and trebled itself—every receiving set purchased added from two to six listeners. . . .

So far these listeners were divided up into five hundred or so little groups, each clustered around a broadcast station, for the stations were small. There now commenced an effort to "increase circulation." It took two forms: first, a movement to increase power, thus covering an area of greater radius, and second, when the audience areas overlapped so that the listener could hear two or more stations, a movement to improve quality and capture his ear by excellence. . . .

It was the ear that was wanted—as many ears as could be gotten, for after the first fad days there was nothing altruistic about broadcasting, except as business had learned to be altruistic as a means to success. It cost a lot of money and took a lot of thought to operate a broadcasting station, and there were not ten men in America who spent that money and energy without some thought of return.

The broadcasters of the day were of three classes: first, those in the business of selling radio receiving apparatus—they broadcast for the direct purpose of increasing sales and building up a new business; second, those, such as newspapers, hotels, stores, and chambers of commerce, who broadcast to spread a name for publicity purposes, calling the desired result the "good-will" of the audience and believing that it would lead to larger business—in short, the advertisers; third, religious and educational institutions, who sought by broadcasting to prosecute their missions in wider areas than hitherto and had the vigor and willingness to spend money to do so.

Of these, the former was doomed to ultimate extinction, since his only purpose was the initial creation of the audience—after that his work was done and the expense borne by others—and the latter to a minor and passive part because of the single-sided character of their activity, which inhibited them from attracting a steady audience (the only one of value), and because they lacked the resources, both monetary and imaginative, which would enable them to compete with the subtle appeal of the commercial broadcaster. They were to learn that, except in rare cases, they could best use radio to further their functions by availing themselves of commercial broadcast facilities, which offered carefully nurtured publics and the necessary experience in showmanship . . . either that or go back to the older means of preaching and teaching. . . .

It was the second—the advertiser—group of broadcasters that was to do the pushing and scrambling and growing, lured by the hope of commercial benefits . . . that in its growth was to use up all the "ether" that could be spared to broadcasting, was then to go into a period of headless competition as it reached for greater and greater audiences until it had all but stifled its own ends . . . and was to emerge at last, disheveled but educated, chastened but triumphant, financially bent, but with daylight ahead.

In the beginning of commercial broadcasting, many of its proponents labored under a fundamental misconception. They believed that the publicity accruing to a single broadcaster would make a station financially profitable or at least self-sustaining if operated in his interest alone. . . .

An analogous situation can be conceived by imagining a state of affairs in which the popular magazine was an unknown quantity. Suppose, then, the idea of getting press circulation through the short-story and the article to be born in somebody's fertile brain—and taken up by commerce. Suppose hotels, department stores and newspapers to preëmpt the field and to issue popular magazines bearing their names, featuring the popular entertainment idea, and carrying no other advertising than the statement at the top of each page: "This magazine is presented to you by the *Bon Ton,* the store where you can get better values." The situation is economically unsound. Failure to use the publication for the presentation of more than one advertiser's message is economic waste, since the quality and appeal necessary to collect an audience large enough to do the "publisher" any good are just as expensive used in his sole behalf as they would be in the behalf of forty enterprises. As this becomes evident, he turns to others for commercial support, and finds himself in a difficult position because of his competitive status in the community. Is he to accept the "advertisements" of rival newspapers, stores, hotels as the case may be? The answer to the whole question is that the undertaking does not belong in his hands at all, but in those of a legitimate publishing company, which has no other commercial purpose than maintaining the magazine and which will serve *all* commerce equally well. . . .

This is precisely the cycle that radio-broadcasting went through, for rather strangely, not one per cent of the original broadcasters entered the game as "publishers." Broadcasting was not conceived to be an independent enterprise—it was begun as an offshoot of merchandising.

The immediate result was that it failed in its purpose. Its cost was greater than the increase in revenue which it brought about. This led to the belief that if the audience could be increased and enough ears reached, enough good-will would be created to balance the expense . . . but unfortunately, experience demonstrated that increasing the station power and improving program quality increased the cost out of all proportion to the increase in audience, so that the books still balanced on the red side. It came to be realized, then, that the broadcasting station must be, in truth, a publishing house,—that the broadcaster must turn to other commercial enterprise with the proposal to sell "time on the air"; . . . in the scramble for revenue that ensued, stations fought for circulation harder than ever. . . .

But before anything else could be done certain engineering problems had to be solved, for with increase in power there came an insufferable condition of interference in the ether abetted by the primitive and relatively non-selective receivers of those days. The audience circles overlapped until instead of creating a larger audience the broadcaster had by his very activity destroyed it for himself and everyone else. . . . Until these engineering problems could be solved, the industry could not hope to stabilize.

In the beginning it had been believed that broadcasting would have no physical limitations—that there was an infinite amount of room on the ether, and furthermore, that by making stations powerful enough, any one of them could be brought to the point of reaching every spot in the country. Both of these beliefs were erroneous. The facts were that, as matters worked out, there were for the time being at least, only about a thousand "channels" or non-interfering wave lengths in the entire radio-wave spectrum of the ether, and that of these only eighty-nine were available for American broadcasting; secondly, that no one station within feasible limits of power could consistently cover the country (the 1928 largest—50,000 watts—though it was on occasion heard over very great areas, had an all-year day and night radius of less than 150 miles); and thirdly, that beyond the limit at which a station's waves were audible, there was an area with radius at least ten times as great, in which the waves interfered with any waves of similar frequency. . . .

When broadcasting began, the only regulatory law covering it was the old Radio Law of 1912. This provided merely that every radio station must be licensed by the Secretary of Commerce, who was to assign to each a wave length that would promote communication and minimize interference. This law was so framed that it gave the Secretary neither power to deny a license nor to enforce the use of the assigned wave length—such contingency was not anticipated in 1912, and as a matter of fact the licensing feature had been chiefly included because of the International Radio Convention which required an exchange of information as to what stations each country had, etc.

In 1921, then, the first broadcasters applied to the Secretary of Commerce, Herbert Hoover, requesting station licenses and wave length assignments. There was, at the time, increasing activity in all the utilizations of the radio art—a simultaneous demand for space on the air not only from these broadcasters but also from various other radio agencies. The Secretary's radio duties were being administered by the Radio Division of the Bureau of Navigation in his department, which picked a broadcast wave length of 360 meters and assigned it to all the first applicants.

By early spring of 1922 there had been so many stations placed on this wave length that, even though they were all using low power, they had begun to interfere with each other. A second wave—400 meters—was taken, then, and a general policy formulated of putting a relatively small number of stations, those with higher power, on this latter wave, and continuing to place all the small ones on 360.

On February 27th, 1922, Secretary Hoover had called an interdepartmental conference to consider the post-war distribution of radio wave lengths; this had been a routine affair of straightening out possible conflicts between Army, Navy, Commerce, Agriculture, etc. A year later, when there were over five hundred broadcasting stations and the first real tendency to go after an audience with higher powers was manifesting itself, he called another.

There was now little doubt that broadcasting was to become an industry, a permanent addition to the nation's communications facilities. To this second conference the Secretary summoned, besides the various departmental representatives, a number of radio engineers, and asked their advice as to the proper allocation of wave lengths for broadcasting. This was in March, 1923, and was the first attempt to reconcile broadcasting's engineering limitations with its economic tendency to spread.

This conference recommended that the old term—wave length—be abolished, and in its stead substituted the designation "frequency"—the number of oscillations of the radio wave per second, expressed in *kilocycles,* or thousands of cycles; it advised that almost all of the band of frequencies between 550 and 1350 kilocycles (wave lengths between 545 and 222 meters) be set aside for broadcasting; it divided the country into five radio zones; it established the engineering principle that to avoid interference, station frequencies must be spaced 10 kilocycles apart and that stations within the same zone must be spaced 50 kilocycles apart; it concurred with the Secretary's classification of stations as Class A, those with less than 500 watts power, Class B, those between 500 and 1000 watts power, Class C, those very small stations content to remain on the old 360 meter wave length (830 kilocycles frequency); applying these general principles to the band of channels reserved for broadcasting, it found that there were some 86 non-interfering frequencies available; it recommended that 44 of them be distributed to the Class B stations, 31 to the Class A, and 1 to the Class C.

These recommendations were carried out, and the resultant situation was, of course, an enormous improvement over the old scheme of things. With 86 channels over which to route the traffic of the 570 broadcasters there seemed room on the air for all the broadcasting activity until kingdom come. The stations were encouraged to increase their power for the benefit of the listeners. They did not need encouragement; as large an audience as possible was vital to every one of them—by September, forty-four of them were using 500 watts and one or two were remodeling their stations to twice that. . . .

But economics was hitting the broadcaster in those days. Try as he would, the cost of station operation would not stay down. His public would not listen to phonograph records or volunteer talent; he faced expense for programs, for station enlargement,—and to add to his financial problem he ran foul of the copyright question. Song-writers were entitled to a royalty on performances of their works for profit; broadcasting was held by the courts to be for profit, and the song-writers' associations demanded an annual royalty fee (from $500 up, depending on the size of the station—in 1928 the largest stations were paying $25,000 a year) from every broadcasting station. For another thing, dozens of broadcasters having quite ignored the question of patent infringement in having their stations built by local and amateur engineers, the American Telephone and Telegraph Company, owners and licensees of the patents covering radiotelephone transmission, now very legitimately demanded that these infringers either pay royalties or stop using the circuits. The Telephone Company was deeply on the debit side of the radio ledger after its many years of development work; in common justice to its stockholders, whose money had thus been spent, and to those who had bought legal transmitting instruments, thereby paying their small share of the development cost, it was morally bound to go after the infringers.

But many a small operator, faced with these copyright and patent royalties (at the time, the Telephone Company license cost from $500 to $2000 a year, depending on the size of the station—churches and schools were licensed for $1.00) found them the straw that broke the camel's back, and went down . . . often taking out upon these seemingly visible antagonists the resentment that should have been directed at an economic situation and at himself for failing to grasp its principles. These were men who would have been far too intelligent to rush into the publishing business without knowledge or preparation, but with that sublime and frequently greedy faith that characterizes all booms they had turned to radio for fame and riches . . . they represented, from an evolutionary standpoint, the "unfittest." . . .

During the remainder of 1923 and 1924 broadcasting continued its economic cycles. Those that dropped out were immediately replaced by newcomers, so that the total number of stations remained as great as ever—about 550—while the audience, in response to continued improvement in programs and receiving apparatus, increased to twenty million. During this period many broadcasters began to offer their services to advertisers, but it was difficult to sell time on the air. Advertisers wanted a large guaranteed audience, which the radio did not yet offer. They were, moreover, frightened off by public antipathy to early advertising attempts. Lacking technique, many of these took the form of so-called "direct advertising"—minus the cloak of amusement—and to this the audience responded with condemnation. . . . There came to be a feeling that the radio was not well adapted to advertising. . . .

But it was. Every station getting "good-will" was advertising. The truth was that the radio could not be used for unvarnished pleas or solicitations, and a few, recognizing this, began to develop a broadcast art that combined entertainment with the desired publicity. It was, after all, a problem in showmanship. . . .

It was discovered, too, during this period, that some of the frequency channels in use were apparently less desirable than others—and that there were a number of unexpected physical aspects to the broadcast problem. Some spots in the country were "dead," others had exceptionally good listening qualities. . . .

There was a continued movement to get as many of the listening ears as possible. The Class B stations were licensed to use 1000 watts power. Most of them had thus far been using only 500; now many of them remodeled their apparatus to the upper limit. The American Telephone and Telegraph Company carried out its first experiments in interconnecting stations by wire; this was the most practical way found yet to increase the audience. That company sent the program of its station, WEAF in New York, to several other broadcasters; it carried out extensive "hook-ups" for the 1924 party conventions which nominated Coolidge and Davis as presidential candidates; it carried out a Defense Day test; ultimately it established a permanent "network" of stations, interconnected by wire lines, and furnished programs to those subscribing to the service . . . it began to sell time on this "net-work" to national advertisers. The Westinghouse Company also interconnected stations, but using a radio wave instead of a wire. . . . By this method programs broadcast from England were picked up in this country by the Radio Corporation of America and rebroadcast. . . . The big radio laboratories were experimenting with transmitters using 50,000 watts power. . . .

The greatest radio audiences were, naturally, in the large cities where the concentration of population was greatest. It was the cities that had the greatest proportional number of stations, the greatest competition for the ear of the radio audience, the greatest amount of interference . . . listeners were beginning to complain. In the seaport cities there was also interference from ship radio, since the ships were still entitled to use frequencies within the broadcast band. . . .

Nevertheless, it seemed that radio was straightening itself out. The tendency toward increase in the number of broadcasters had fallen off; in September, 1924, there were only 533—six months earlier there had been nearly 600. Congestion on the air, however, had not lessened. The tendency toward increase in power made frequency allocations increasingly difficult, and now all the channels in the broadcasting band were used up. In October, 1924, Secretary Hoover called a third conference to consider the situation and make recommendations.

This conference included among its numbers not only scientists, but representatives of the radio manufacturers and of the broadcasters—the radio industry and the Department of Commerce had always worked hand in hand. In spite of the fact that the Secretary's lack of ability to enforce his assignments of power limits and frequencies was known, the broadcasters had always willingly submitted to his dicta, which had so far tended to clarify and improve the situation—the most arduous thing required had been the division of time between conflicting stations in the cities; even this had been accepted as unavoidable and better than losing the audience altogether as the result of interference.

The conference recommended that the ship-traffic be got out of the broadcast band altogether; that the broadcast band be again extended, this time to include all frequencies between 550 and 1500 kilocycles; that the number of zones be increased to six, with the Pacific Coast using the same frequencies as those assigned to the North Atlantic states; that in general increases in power, even above 1000 watts, be encouraged because they meant better programs through static and a larger permanent radius. . . .

With the adoption of this report there were eighty-nine channels in the broadcast band (several having been allotted exclusively to Canada by mutual agreement)—and by duplicating those on the West and East coasts this was some thirty more than had hitherto been available. Once more expansion had been the solution; once more there was room on the air for everybody.

But it was only a temporary respite. By the end of another year the number of stations equipped to use 500 watts or more had increased from 115 to 197; on the earlier date there had been only 2 using more than 500 watts—there were now 59, of which 32 were using 1000 watts; 25, 5000 watts; and 2 still higher powers! Once more the broadcasters had crowded their portion of the ether so full that the radio highways were jammed.

Always before in such circumstances it had been possible to widen the road. Now the limit had been reached. Not only were other legitimate radio activities in need of space on the air, but the five or six million home receiving sets in the country were not built to receive beyond the 550-1500 kilocycle band, so that any extension of the broadcast spectrum would do no good.

For the fourth time Secretary Hoover called a

conference, and once more—in November, 1925—the representatives of the industry assembled in Washington with great protestation of willingness to work together to solve the problem. The Secretary laid the matter squarely before them, and at last the fell words were spoken—the industry would have to cut down its activities; it had run up against natural barriers. Not only was it already overflowing its ethereal banks, but there were on hand 175 applications for new licenses.

The situation had its element of humor. The voice of the conference was the voice of those who had always hitherto gotten what they wanted. They were themselves entirely willing to coöperate—mutually and with the Secretary—but they were unduly optimistic when they attempted to do the coöperating of the gentlemen on the outside looking in. They recommended that the Secretary issue no more licenses until the number of broadcasters had decreased by natural mortality. . . .

Yes, there were too many broadcasters, and everybody knew it. Everybody knew, too, that as far as the Secretary was concerned, he had no power to deny licenses—the 1912 law had no teeth in it. Rather the hope of the conference, in repeatedly recommending that he curb the bucking situation, was that legislation could be got through Congress that would give him the needed authority.

That winter a new Act to Regulate Radio Communications was brought before the House by Representative Wallace White of Maine . . . but it stranded on the rocks of Congressional unwillingness to give the control of such an extensive medium of audience-appeal into the hands of any one man or Department . . . and so at the crucial moment when the irresistible force was about to meet the immovable body there was nothing to say it nay. . . .

Everybody had been waiting to see what would happen in Washington. Now the storm broke. A station operated by the Zenith Radio Company in Chicago—Chicago was perhaps the most congested radio crossroads in the country—led the way and jumped to a Canadian frequency, increasing power at the same time and thus making use of both the factors needed for larger audience-getting. Another, in Shreveport, Louisiana, which had not been receiving what it thought its radio due, promptly followed suit. When the Secretary of Commerce attempted to deny the Chicago station a renewal of its license it took the matter to court, and the courts decided that, as had always been known and feared, the Secretary lacked authority. The license was issued and all control had broken down.

This was in July, 1926. There now ensued a period of comparative chaos. One hundred and fifty-five new stations took the air, bringing the already overlarge total to a new peak beyond the seven hundred mark! The frequency allocations that had been so painfully worked out went by the board as those assigned to the most progressive stations were encroached on by one after another of those who wanted audiences without earning them. Coupled with this frequency-jumping was a simultaneous defensive tendency to increase power for the purpose of drowning out the invaders and forcing programs through to the largest audience possible. . . . There was a general conviction that hard and fast regulation would come soon, now; as much as anything this chaotic scramble was an attempt to reach a good strategic position in the hope that possession would be nine points of the law. . . .

Meantime the more stable broadcasters, resigned to the moment and sure that in the long run they would get their just deserts and regain their privileged places on the air, were paying much attention to the economic angle of program quality. Program quality had been a factor of constantly increasing expense, for broadcast technique had evolved itself into a complex mixture of the vaudeville show, the recital, the band concert, the lecture and the symphony, with whichever preponderant that would appeal to the precise audience desired. The proper presentation of these diversified elements had developed a new variant of showmanship, in which the country's amusement centers had led because of the superior entertainment and showman talent available there. Indeed, there was coming to be a natural dearth of radio artists and program directors in the midland towns, due to the tendency of those who were successful there to gravitate toward New York, Chicago, Los Angeles, where there was more radio activity, greater return. To hold them, they must be given metropolitan money . . . the cost of operating a properly competitive station had risen to the neighborhood of $100,000 a year—such a station cost at least $150,000 to install. . . .

So far the happiest solution of this economic situation had been the Telephone Company's network of interconnected stations, which now had eighteen "subscribers" and was working out very well—it made it possible for stations in the interior to broadcast, for far less money than if they had put them on themselves, the same good programs that were available to New York audiences; it made it possible to guarantee an advertiser a large "circulation."

This activity was outside the Telephone Company's normal field of operations; it had been entered into for experimental purposes and as a means of stimulating the use of wire lines—and though successful to the point of presaging a good future, it had cost a great deal of money which there was no immediate prospect of regaining. The idea was now taken up by the Radio Corporation and its affiliated manufacturers, who were most interested in seeing broadcasting placed on a stable economic footing . . . (the Telephone Company had sold its stock in the Radio Corporation and was no longer represented on the board of directors, though its patent cross-licensing arrangement still continued in force). . . .

In November, 1926, then, the Radio Corporation of America purchased, for $1,000,000, station WEAF in New York (thereby retiring the Telephone Company from active participation in broadcasting activities) and incorporated that station under the name of the National Broadcasting Company to carry on the chain broadcast idea and place it, if possible, on a sound financial basis.

The National Broadcasting Company furnished "sustaining" programs to its subscriber-stations for $45 an hour. By agreement with them, it offered its facilities to advertisers at rates based on (1) the cost of the program and talent (which the advertiser might furnish himself if he desired, but which in that case must come up to rigid standards), and (2) the cost of the wire lines and of the use of the subscriber-stations. The latter part of the cost depended, of course, upon the number of stations in the "hook-up." When taking "sponsored" programs, each subscriber-station received $50 an hour as its share of the returns. . . .

Several phases of this company's operation were indicative of the advanced business thought of the day—which recognized that all business was a public utility—and of the serious consideration which was being given to the improvement of radio programs. Its services were open to anyone with the money to pay for them, including radio-manufacturers in competition with its owners. It had an advisory council (non-remunerated except for traveling expenses to and from meetings) which included such names as Charles Evans Hughes, William Green, Elihu Root, Dwight Morrow, John W. Davis. It was a tremendous financial venture—and one whose economic justification only time could prove or disprove; it lost $500,000 during its first year of operation....

Curiously enough, the radio now became the largest user of telephone lines in the United States! During 1927 the National Broadcasting Company paid $1,600,000 in tolls for the specially high quality circuits that it used!

When the sixty-ninth Congress of the United States met for its second session it was faced with the urgent necessity of doing something about Radio. The White Act, up again for consideration, proposed that control be given to the Secretary of Commerce, with an advisory commission of nine . . . but Congress, as ever, would not consider any such proposition. Instead, the Radio Law passed by that body and approved on February 23rd, 1927, created a Federal Radio Commission of five men—one each from five newly created radio zones covering the country—to which was assigned the task of straightening out the radio muddle in one year, after which the administration of their solution was to be left to the Secretary of Commerce. The Commission was thenceforward to serve as a radio court of appeals for the settlement of any conflicts that might arise due to the Secretary's decisions. . . .

Now, at last, the unlimited growth of broadcasting activities was stopped. Whatever else the law did or did not do, it provided a body that had the power to withhold a license if it believed the public interest, convenience, or necessity did not warrant granting it. . . .

The Radio Boom was over.

To consider the effect of the Radio Boom upon the radio-manufacturing industry, one must go back again to 1921 before the boom was born. In that year the sales of radio apparatus in the United States amounted to about nine million dollars. In 1923 they were forty-six million; in 1926 four hundred million dollars.

These are figures indicative of the vigor of post-war America, a nation that had scarcely paused in its stride for the absorption of the war upheaval . . . figures indicative of the way it was rushing on to its golden century . . . of the stabilizing that came into its life after those first post-war years when everything was an enigma, normalcy a dream. . . .

And they tell of the growth of a major industry.

Apart from the big electrical manufacturers who had interested themselves in the radio art for communications purposes, there were in 1921 some two dozen small companies making "parts" and the like for radio amateurs. Most of these had been licensed by Edwin Armstrong to use his "feed-back" patent before he sold it to the Westinghouse Company . . . they still retained that right, but neither they nor the big companies were destined to provide the receiving apparatus that went into American homes by the hundreds of thousands during the early part of the boom. The demand was far greater than the industry.

To fill it, hundreds of set-makers — they scarcely warranted the designation "manufacturer" —sprang into being overnight. Electricians, war-time radio operators, amateurs . . . they did not have to seek trade—their friends came to them. Some built one set, some two, some went into business. Naval officers resigned from the service . . . engineers left their laboratories . . . it seemed that a flood of dollars would pour itself into the pockets of any man who could make radio apparatus. . . .

At first it was crystal sets they made; afterwards they used the latest circuits upon which they could lay their hands. When, in 1922, Armstrong invented the *super-heterodyne* (and sold it to Westinghouse, General Electric and the Radio Corporation under an agreement whereby they had had the right to an option on all his future patents) and L. A. Hazeltine of Stevens Institute invented another non-radiating circuit called the *neutrodyne* (and licensed it to various manufacturers) . . . they used both those circuits. . . .

But these men could not long hold the radio-manufacturing field, for making sets was an art, and business was business. There inevitably appeared organizations more soundly conceived, amply financed and able to produce a product beside which the attic-made affair was a clumsy crudity.

Even these had their difficulties. They could not, like the fly-by-nights, infringe patents—unless, then, they secured licenses from the owners of the radio patents, they were unquestionably powerless to manufacture legally any modern tube receiving-apparatus. . . . Further than that, unless they were in touch with laboratories like those in Schenectady, Pittsburgh, and New York, they were working in the dark, for those laboratories were months in advance of the times . . . the manufacturer who could not plan his production on the basis of what was coming was in danger of finding his entire product obsolete overnight. That happened to some, and they went bankrupt; hundreds of sets were thus thrown on the market at a tenth what it had cost to make them, thereby adding to the woes of those who were trying to stabilize the industry.

As for the Radio Corporation, that enterprise found itself in a situation that would have been ludicrous had it not been so serious. Conceived as a communications enterprise and engaged in carrying out its mission of establishing world-wide wireless with all the vigor at its command, it found that this new broadcast thing, which had at first seemed rather unimportant, was now dwarfing, financially, the communications end of its activities! Here was, in effect, the old American Marconi Company, which had for years lived for nothing but ship-shore and overseas telegraphy, turning into a sales agency for a form of parlor amusement! It was enough to make an old communications man roll in his grave!

And now, just as it had after all these years gotten sufficient fundamental patents together to enable it to communicate properly . . . it found that the possession of those patents made it an object of suspicion and distrust to the growing industry! "Monopoly!" "Trust!" The whole thing, said certain voices, had been a deep-laid plot! The Radio Corporation was a stifler of industry . . . it had known all along that this was coming; it had cornered the market in radio devices and it was going to keep them away from everybody . . . it was the antagonist of the entire nation. . . .

The serious aspect of the matter was that in 1923 the Congress pricked up its ears and demanded that the Federal Trade Commission investigate radio manufacture with a view to ascertaining whether or not it was being monopolized, and that in 1924 the latter body charged the Radio Corporation *et al.* with violation of the anti-trust laws and instituted one of those trials . . . four years later six hundred thousand words of government testimony had been taken, the "exhibits" filled a truck, and the defense of the eight "conspirators" had not even begun. . . .

The humorous aspect of the matter was that the Radio Corporation was as much at sea as everyone else as to what the whole radio boom was going to lead to, that above all things it desired stability and an end to the middle . . . that in 1923 it would have given half a dozen patents to know what the exact situation would be in 1930. . . .

And the practical aspect of the matter was that, whatever rights it did or did not have to monopolize radio by virtue of its very real combination of patent monopolies, the Radio Corporation did not monopolize anything, nor could it. The radio boom had not asked anyone's permission to take place. . . .

But when it came to large-scale manufacture, the Radio Corporation could very righteously step in to prevent anyone from filching those devices and circuits that had cost so many millions . . . and using them for profit. Approached under these circumstances by the twenty-five odd radio manufacturers who had emerged from the heap as substantial enterprises, the Radio Corporation granted them license to use all its radio patents except the *superheterodyne*. As a royalty fee, it charged seven and a half per cent of the wholesale price (which they passed on to the buyer, where it belonged); this was applied to writing off the sums expended in patent development. And in the interests of keeping the industry in the hands of stable people, it made a provision that this royalty was to be not less than $100,000 a year.

Having thus literally created twenty-five ardent competitors, it rolled up its shirt-sleeves and began to scramble for business.

The tube manufacturing situation, however, was a different matter. The modern tube was purely a General Electric and Telephone Company development, and it was proposed to retain its manufacture strictly in the hands of its creators and their cross-licensee, the Westinghouse Company. . . .

This decision led to a pretty squabble on the part of those known as "bootleggers" of tubes. The bootleggers had appeared as far back as 1915, and in numbers in 1920. A cloak of legitimacy had been lent to their operations by the expiration of the Fleming Valve patent in 1922, and later by the expiration of the Audion patent in 1924. But none of these people wanted to make either Fleming Valves or De Forest Audions; they wanted to make modern pure electron discharge tubes . . . and none of them would the Radio Corporation license. Cunningham, true, had been admitted, but only as a sales agent . . . the tubes he sold were Radio Corporation tubes under his name. The others, left in the cold, retaliated by sniping from every corner. They banded together to form a "Protective Association," and for a group who wished to do a manifestly illegal piece of infringing, they exhibited surprising activity. Their Washington lobbyists made the welkin ring with the "trust" cry; they gloated over the Federal Trade Commission investigation; they even succeeded—in the hope that the Radio Corporation would be dissolved—in getting a bill introduced before Congress which contained the rather remarkable provision that in the event any holder of United States patents were found guilty of violating anti-trust or "restraint of trade" laws, he should lose all rights to enforce those patents! The bill died in Committee, but as a sample. . . .

Nor was the Radio Corporation quiescent under this active attack. It lobbied a bit itself. Advised by its lawyers that it was within its rights in doing so, it included in the license-contracts with its twenty-five competitors a provision that their sets as initially manufactured should be equipped with Radio Corporation tubes (since the tubes were part of the circuits and the licensor had the right to require that he furnish any part of a licensed device); the bootleggers counter-attacked by taking this provision to court and getting an injunction prohibiting it; the Radio Corporation appealed. . . .

It was as good as a play, but the humor was sardonic for no matter who won, the bill was the public's . . . legal costs were added in as "overhead" —had there not been so many suits, tubes might have been five cents cheaper! As it was, they started at seven dollars in 1922 and had dropped to less than a third of that in 1927 . . . for which the bootleggers took the credit, saying that if it had not been for them, the Radio Corporation would have charged seven dollars forever. . . .

Business remains a form of warfare, even in the scientific era. . . .

Meantime, undisturbed by booms or business, the research laboratories forged surely and steadily ahead. The list of their achievements during this period read like a catalogue. They took a loudspeaker that was in 1921 a trial to listen to, and made it an instrument that was a marvel of reproductive quality. They made it possible to build radio receiving sets, delicate as watches, by quantity production methods. With tubes they did the impossible over and over again—made them first to operate with dry batteries (by mixing Thorium with the Tungsten of their filaments; when they were heated the Thorium worked its way to the surface and formed a layer *one atom deep*—it gave off a much higher thermionic emission than Tungsten, but could not be made into a filament by itself) and incidentally revived a nearly extinct dry battery industry; made them to operate on alternating current; made "screen-grid" tubes, rectifier tubes, power tubes for operating dynamic speakers, tubes with dozens of invisible internal improvements, hundred kilowatt tubes for transmitters, tubes, tubes, tubes. . . .

Transmitter design changed so rapidly that a transmitter a year old was obsolete . . . they could do anything with oscillating circuits—pluck harmonics out of them, amplify them, transform them, fling their vibrations into space and snatch them out again . . . and always retain the subtle shading of tone modulation that made the result an uncannily faithful reproduction of the source.

The home radio set was disguised as a piece of furniture; its three dials were cut to two, to one . . . it was refined and refined again . . . its circuits were tuned to a sensitiveness that made them able to pick up a very whisper of a Hertz wave and send it thundering out like the voice of a mountain. . . .

The laboratories remained the well-spring, the key-stone of the scientific era . . . and all over America the radio was grown to be as much a part of the scheme of things as was the daily news.

A year after its conquest of America, the radio boom spread abroad. It found the nations of Europe still militarized and engaged in many post-war suspicions and distrusts—radio in almost all of them was yet under government aegis, and broadcasting, when it began, was made a government function because of the old fears of its potential powers over the opinions of its audiences.

To cover its cost a general policy was adopted requiring the set-owner to pay an annual tax. In England this was ten shillings (about $2.50), and the money so collected was turned over to the British Broadcasting Company—a private company to which had been given a contract to erect and operate broadcasting stations for the state. The programs sent out, then, contained no advertising, and a strict censorship was maintained over speeches, etc.

There were many who proposed that a like system be adopted in the United States. It had the fundamental weakness, however, of being economically unsound—it wasted the enormous value of an audience. Broadcasting as it grew up in America brought more than a direct benefit to the public; it stimulated trade, just as did the newspapers and magazines; it promoted prosperity by keeping money loose and aiding business. Had all American advertising been stopped, the nation would have suffered "hard times" within a month; had broadcast advertising been stopped, the trend would have been similar . . . a valuable medium would have been deprived of one of its greatest powers to do good.

The reason for such a proposal was an aesthetic one. Broadcast advertising as practiced in the early days was crude, blatant, an offense to taste. But business was too resourceful to allow such a condition to continue, and by the end of the radio boom the aesthetes' plaints were stilled . . . voices that it had once been the privilege of the few to hear now flew to the ends of the continent; radio programs were "reviewed" in the press; Mr. Graham McNamee's name was a household word; millions went to concerts, to football games, to receptions to aviators . . . and the aesthetes were among them.

CHAPTER ELEVEN

ETHEREALLY LINKED CONTINENTS

WHILE the radio art had been making its way into every man's home, it had also been proving itself a public servant of value equal to the cable as its communications circuits spread over sea after sea, binding far-flung nations to each other by ties of daily intercourse. Through it, America, now busily engaged in commercial trafficking with every nation on earth, had for the first time a comprehensive international communications system of her own . . . and she used it; in 1921, 7,001,610 paid words passed through her radio stations—in 1927 the number had increased to 38,662,536. Reduced transoceanic communications rates had saved the country thirty million dollars.

These international circuits had been established along different ideas of radio-communication than those originally propounded by Marconi and his associates, who had held that the stations at each end of a circuit should belong to the same company. What moves had been made in that direction after the war, however, had been checked by (1) lack of traffic, (2) scarcity of wave lengths, and (3) a universal tendency on the part of national governments to prohibit within their borders radio stations owned by other than their own nationals.

The basis of the solution was economic. Suppose, for example, that the United States and Germany wished to communicate with each other. This could be done by radio if (1) an American company erected adequate stations in both countries (thus using two wave lengths), or (2) if a German company erected adequate stations in both countries (also using two wave lengths), or (3) if German and American companies each erected stations in both countries and competed for the traffic—this would use up four wave lengths, and each company would normally get only that business originating in its own country—or (4) if each company erected a station in its own territory and they agreed to exchange traffic—this would result in a competitive situation, as each would act as a check upon the other; it would make full use of the facilities, and it would require only two wave lengths. . . .

It came to be seen that so far as radio was concerned, competition by duplication of facilities was the death of trade—it was inefficient and uneconomic. . . .

In place, then, of a number of separated transoceanic radio enterprises, there grew up a network of coöperating international stations, some of them privately owned, some of them owned by governments. The result of this coöperation was a revelation of the benefits to be derived from international teamwork, and an indication that the lessons of the war had taken root—it was a thing that would have been impossible in 1912.

The "business treaty" that now made such coöperation possible was the Traffic Agreement—a mutual contract for the exchange of messages between two radio-communications agencies. In some cases this contract was exclusive; sometimes it was not. An integral part of it, usually, was an agreement to exchange information, to pool knowledge, to make devices available . . . it was a splendid arrangement—in effect it gave each nation an ethereal voice and signified her willingness to use it.

For the United States, the Radio Corporation was from the moment of its formation the active agent for commercial radio-communications. Between 1920 and 1928 it signed Traffic Agreements with the British Marconi Company, the French *Compagnie Radio France,* the Minister of Posts of the German *Reich,* the Imperial Japanese Government, the Director General of Telegraphs of the Government of Norway, the Italian Government and *Italia Radio,* the Ministry of Posts and Telegraphs of the Republic of Poland, the *Trans Radio Argentina* Company, the Royal Telegraph Administration of the Kingdom of Sweden, the Colonial Government of the Netherlands East Indies (Java), the Brazilian National Company, the French *Compagnie Générale de Telegraphie sans Fil* (for a French Indo-China circuit), the Netherlands Administration of Posts and Telegraphs, the Administration of Telegraphs and Telephones of the Kingdom of Belgium, the Republic of Turkey, the *Companhia Portuguesa Radio Marconi*, the Government of Liberia . . . and apart from these opened circuits through its own stations or by other means with Honolulu, the Philippines, St. Martins in the Dutch West Indies, Dutch Guiana, Colombia, Venezuela, Porto Rico, Curacao, Hongkong, Shanghai. . . .

Communications were begun as soon as adequate stations were ready; the rising tide of radio's use for overseas telegraphy was marked by the dates of opening of the Radio Corporation's circuits:

To British Marconi	March	1,	1920
Hawaii and Japan	March	1,	"
Norway	May	17,	"
Germany	August	1,	"
France	December	14,	"
Italy	August	10,	1923
Poland	October	4,	"
Argentina	January	25,	1924
Sweden	December	1,	"
Java	July	16,	1925
Brazil	May	3,	1926
French Indo-China	September	15,	"
Holland	November	1,	"
St. Martins	June	21,	1927
Philippines	June	27,	"

Dutch Guiana	August	9,	1927
Columbia	August	12,	"
Venezuela	August	18,	"
Belgium	October	3,	"
Porto Rico	October	10,	"
Hongkong via Manila	October	18,	"
Turkey	December	10,	"
Shanghai via Manila	February	21,	1928
Portugal	April	2,	"
Curaçao	August	4,	"

All of these circuits except those to South America and China were established without particular difficulty other than that normally incident to business negotiations. In the latter countries, however, the national governments were neither sufficiently prosperous nor sufficiently advanced to inaugurate external communications of their own—they were in somewhat similar position in that respect to the United States of the nineteenth century; other nations were much more interested in maintaining communication with them than they themselves were in maintaining communication with the outside world. Consequently, he who would open radio circuits to them must build stations of his own.

After the war, England, France, Germany, and the United States were all four looking to South America as a source of raw materials and as a market. The trend toward radio-communications inevitably followed, but when the Radio Corporation investigated the South American field it found that it had been anticipated by the *Compagnie Générale de*

Telegraphie sans Fil, the German *Trans Radio,* and British Marconi, all three of whom had concessions in one or another South American country, and some of whom already had stations in operation.

For obvious reasons it would not do to make traffic agreements with these companies; the very reason the Radio Corporation had been formed was to enable America to communicate in such situations without depending upon the facilities of her trade competitors.

Nor was the demand for communication to South America great enough to support four competing radio agencies. As the directors of the Radio Corporation later reported to their stockholders: "The erection of individual stations by different nationals would have meant duplication of capital in countries where the prospective business was too meager to warrant such duplication, particularly as the construction of stations was very expensive, and the wave lengths suitable for long-distance international radio communications were so few that they should be used at their full capacity; moreover, the national feeling with reference to communications ran too high to permit the successful execution of competitive programs. To have proceeded with individual competitive stations would have been highly wasteful and uneconomic."

Since this state of affairs was apparent to the other companies as well, representatives of all four met in Paris in October, 1921, and after deliberating, agreed upon a solution along coöperative lines. The agreement which the four companies then signed created a Board of Control, to handle their combined South American radio-communications activities.

This Board was to be composed of nine men, two representing each company, the ninth and chairman to be the additional appointee of the American Company and to be an American of high standing not connected with that company—his presence on the Board guaranteed the "Monroe Doctrine of Radio" upon which the Radio Corporation had insisted. The chairman was given a vote in case of tie, and the agreement further provided that: "In any case in which the voting . . . shall so result as to leave the trustees appointed by any one of the four parties in a minority and the said party believes that injustice has been done to it . . . the trustees appointed by the said party may appeal to the chairman to veto the proposed action of the majority; and if the chairman, after full discussion, is of the opinion that the said proposed action would do substantial injustice to the said minority, he shall have the power to forbid it and to cancel the resolution agreed to by the said majority."

Acting in its trustee capacity, the Board was to form subsidiary radio-communications companies in each of the South American republics to which it was deemed desirable to extend communications. These were to be known as National Companies; at least sixty per cent of their voting power was to be controlled by the trustees; they were to be managed by boards of directors representing the four owner companies; their Presidents were to be citizens of the country in which they were incorporated . . . as were preferably their active Managers also.

Under this plan operations went ahead rapidly. The existing stations in Brazil and the Argentine were taken over by the Board, National Companies formed, and the needed communications circuits opened . . . South America, North America, and Europe were linked, then, by radio. . . .

The Chinese puzzle, however, did not offer such a simple solution.

The original American post-war venture into the Chinese radio field was made by the Federal Telegraph Company of California. As that company had prepared to resume operations after the war, it had seemed futile to recommence trans-Pacific communication activities unless a California-China circuit could be inaugurated. The Radio Corporation had taken over the Marconi Company's old Traffic Agreement with Japan, and was operating to Japan and Hawaii; there was cable service to Hawaii and the Philippines (operated by the Pacific Commercial Cable Company, 75 per cent of whose stock was British owned)—only a circuit to China, particularly if an exclusive Traffic Agreement could be gotten, would make Federal operations on the Pacific thoroughly worth while.

To see whether or not such a thing was possible, a Federal representative was sent to China in 1920, to negotiate with the government of the Republic of China.

He was successful. On January 8th, 1921, a contract (supplemented by an agreement dated September 19th, 1921) was signed with that government, which provided for the erection by Federal of a duplex super-power (1000 kilowatt Federal arc) station at Shanghai for transoceanic communication, and for the erection of four smaller stations at Harbin, Peking, Canton, and Shanghai, for communication within the republic. These stations were to be built at an estimated cost of $13,000,000, each party contributing half that sum, and the Chinese government was to deliver immediately to the Federal Company $6,500,000 in 8 per cent Chinese Government bonds (countersigned by the Chinese Ambassador in Washington) as its share in the venture, these to be marketed by the company as necessary to secure working capital, but to remain, of course, obligations upon the Chinese Government. Under the terms of the contract the stations in China were to be jointly operated for ten years by a body known as the China-Federal Radio Administration; at the end of that time it was estimated that the revenues would have retired the bonds and that the Chinese operating personnel would have become sufficiently expert to take over the management—Federal would then retire, but would continue to receive royalties for another ten years. It was provided that the China-Federal Radio Administration would route all its messages for America via its radio associate in the United States, for twenty years from the date of completion of the last station to be erected.

The United States State Department had lent diplomatic aid in the negotiations that led to the signing of this contract; that department now became, for the sake of America's "face," extremely anxious to have the project consummated as pronounced opposition to it, on the part of other nations upon the Chinese government. Under the Door" policy America's stand was that no nation should be given monopolistic privileges there that would deny trade opportunities to other nations... but as news of the Federal-China contract got abroad, it developed that there were at least four agencies, involving three different countries, which claimed exclusive rights to all communications to and from China!

There was (1) the Great Northern Telegraph Company, a Danish cable concern, which claimed an out-and-out communications monopoly under an 1896 contract with the Chinese government; this contract did not expire until 1930. There was (2) a contract in existence between the British government and the Chinese government which called for the construction of certain land and cable telegraph lines, and which had certain clauses guaranteeing protection against competition. There was (3) an existing radio company known as the Chinese National Wireless Telegraph Company, owned half by British Marconi and half by the Chinese government; it had been formed to manufacture radio apparatus for the Chinese Army, and now maintained that it had monopoly rights to the manufacture of all radio apparatus for that government. And last of all, there was (4) the Japanese *Mitsui Bussan Kaisha,* which had a radio station in Peking . . . it now developed that in connection with the right to build that station, the Japanese had secretly been given a monopoly of all radio-communications to and from China for thirty years!

As the Danish, British, and Japanese governments began to press, through diplomatic channels, these "monopolies" granted to their nationals by the beneficent Chinese, the United States State Department took up the cudgels in behalf of Federal, but to no avail—the American company found itself blocked at every turn. The bonds—essential for commencing operations—had been printed and turned over to the Chinese Ambassador for signature; that dignitary signed them, but at the last moment refused to deliver them, raising some difficulty having to do with the coupons that had come due since the bonds were authorized. Strangely enough, he sailed for Europe on the same day; attempts to gain delivery of the bonds after his return met with the same baffling evasions . . . as a matter of fact they were never delivered at all.

Meantime there remained the contract, perfectly valid—but in the spring of 1922 President R. P. Schwerin of the Federal Company felt that he had taken the matter as far as he or that company could carry it alone, and as his next step sought the aid of the Radio Corporation in the hope that through its greater strength and influence in the radio field, it might be able to accomplish what he was finding impossible.

The Radio Corporation had been watching the situation—which was, in effect, the same sort of deadlock as had existed in South America prior to the formation of the Board of Trustees—since its inception. Conceiving it to be the Corporation's duty to further the extension of American communications in every way possible, it had, through Owen D. Young, chairman of the board of directors, addressed letters on December 7th, 1921, December 12th, 1921, and January 9th, 1922, to James R. Sheffield, Elihu Root, and Charles E. Hughes, delegates to the Disarmament Conference then meeting in Washington (and on whose agenda was the subject of electrical communications in the Pacific). These letters had outlined the situation, and suggested international coöperation through a board of trustees as the solution in China if speedy arrival at communications service was the end desired.

In this connection the Radio Corporation had

offered to coöperate with the Federal Company if terms satisfactory to both could be arrived at, and if Federal would agree that the Radio Corporation should handle the American end of the Pacific circuit. The Radio Corporation had never believed in the existence of a second trans-oceanic radio company on the Pacific Coast, not because it wanted a monopoly, but because it had learned in the past two years' experience that when competing American radio companies went to a single foreign radio agency, such as the Japanese government, seeking Traffic Agreements, it was the foreigner who held the strategic position, who dictated the terms, who established the rates. . . .

Elihu Root replied to Mr. Young that he was convinced it was a project beyond the scope of the Conference, and one that should be taken up through the ordinary channels of diplomacy. Nevertheless, the Conference did consider the matter in committee, and recommended the coöperative idea as its solution.

When Mr. Schwerin approached the Radio Corporation, then, immediately after the Conference, there were two alternative courses for the achievement of communications with China: (1) an attempt to arrive at a consortum between the various nationals holding conflicting concessions there, or (2) an attempt to carry out the Federal-China contract.

Of these, the latter was much the least promising,

in the light of the sentiment of the Disarmament Conference—but it became evident that western efficiency would have to bow to oriental psychology, for the State Department decided that to abandon the Federal-China contract would mean very serious "loss of face,"—would make it seem to the Chinese that the United States had given way to its rivals. . . . The State Department had learned that East was East. . . .

In September, 1922, then, a new company was formed, sanctioned by and with the coöperation of the United States Departments of State, Justice, Commerce, and the Navy. It was called the Federal Telegraph Company of Delaware; it was capitalized for $9,500,000, and bonds to the value of $3,500,000 were issued; it was jointly owned by the Radio Corporation and Federal of California, and to it were assigned the Federal-China contracts. This not only gave the Federal Company the active financial support of the Radio Corporation, but the use of the patents covering the latest developments of the art—it meant that the Chinese stations which could now be erected would be far superior to the stations which the Federal Company would otherwise have been able to build.

In August Mr. Schwerin had sailed for China to secure consent to this assignment of the contracts. He found on his arrival there, that the other nationals involved in communications had apparently foresworn their own conflicts and banded together to prevent any action favorable to him; for nearly a year, aided in every way possible by American diplomatic representatives, he negotiated with the Chinese officials, finding his way continually blocked by most bitter opposition. Only on July 13th, 1923, was he able, at last, to obtain consent to the contracts' assignment. . . .

It never did any good—the muddle could not be straightened out. Neither the land for the stations, nor the bonds covering the Chinese portion of their cost, were ever turned over. . . .

The American parties desired to make Shanghai the "Radio Central" of the Orient, but through no fault of their own they did not succeed. The Radio Corporation of America, despairing of the Chinese project, at length secured a radio concession from the Philippine Legislature and erected its Oriental "Radio Central" there. In June, 1927, communication was begun between this station and the stations in California; messages for China were relayed by radio to Hongkong beginning October 18th of that year, and to Shanghai after February 21st, 1928 . . . the rates were 16 2/3 per cent lower than those of the cable, which met the cut within twenty-four hours. . . .

On August 24th, 1927, the Federal Telegraph Company of California was sold to the Commercial Cable Company interests . . . the Federal Telegraph Company of Delaware held the dormant Chinese contracts and nothing else.

Meantime direct radio service had been established by the Radio Corporation, between California, the Dutch East Indies, and French Indo-China. . . .

For years "World Wide Wireless" had been the slogan of the old American Marconi Company. That slogan had been taken over by the Radio Corporation—now, at last, it had been achieved.

In Britain, radio-communications had not gone forward so blithely after the war, due to the devastating effect of a non-comprehensive government radio policy. . . . As a result, when 1927 dawned, though the British Marconi Company had by Traffic Agreements linked England with most of the foreign nations of the earth, those portions of the globe most vital to her—her dominions—remained poorly served by radio or not at all. To be sure, a better day was at hand . . . but due to her inability to choose between government ownership and private enterprise for handling her inter-dominion commercial traffic she had wasted large sums of money and a great deal of time. . . .

In 1919, directly after the war, the old All Red Chain idea had been got out of storage and dusted off . . . and the stations in England and Egypt, work on which had been suspended since 1914, had been ordered completed.

In June, 1920, an Imperial Wireless Telegraph Committee had confirmed the old idea—stations two thousand miles apart, government owned, to link the various dominions together. This had been in repudiation of a Marconi Company proposal recommending super-power stations for direct communication with India, Australia, and South Africa.

Following the report of this Committee, the Wireless Commission called for in that report, was appointed; this latter body recommended Vacuum Tube Transmitters for seven of the All Red Chain stations, and that the eighth be equipped both with Vacuum Tube Transmitters and with the Arc; the estimated cost of the whole chain was six million dollars. This was in January, 1922 . . . a year and a half had gone by, deciding on things. . . .

And neither the "chain" plan nor its slow progress met with any great degree of enthusiasm on the part of the dominions, which were coming to believe more and more in the direct communication idea. Led by Australia, then, they took action of their own. In March, 1922, that commonwealth entered into an agreement with a Marconi subsidiary known as Amalgamated Wireless (Australasia) (Ltd.) for the erection and operation of a super-power station capable of direct communication with England and Canada, and a system of smaller stations for internal and ship-service telegraphy; the whole project was to be completed within two years, and in furtherance of the basic idea of government ownership, the commonwealth bought a majority of the stock in this company.

South Africa followed suit, and in September, 1922, and March, 1923, arranged with the Marconi Company for the formation of a similar subsidiary and the erection of a super-power South African station. This, however, was to be a Marconi enterprise (with certain features of South African control) for a period of ten years, when the government had the option of purchasing the station.

India, too, made moves toward stations of her own for direct communication with England, while in Canada the Canadian Marconi Company secured licenses for super-power stations at Montreal and Vancouver.

All these plans and activities led the home government to reconsider the matter, and in July, 1922, it approved the Marconi plan for stations in England comparable to those contemplated in the dominions, and announced the abandonment of several of the intermediary links on the Chain. . . .

In 1923 there was yet no All Red Chain, and a new Conservative government decided that matters might be speeded up by admitting private enterprise to the inter-dominion field; an announcement that this would now be permitted was made . . . but it was also stated that the government would continue to operate stations capable of inter-dominion service and would admit them to commercial traffic. . . .

The Marconi Company promptly applied to the Postmaster General for a broad license, seeking sufficient privileges in the way of exclusive rights to justify the enormous investment that would be involved. This application was rejected on the grounds that it would constitute a virtual monopoly . . . the negotiations dragged along for another year during which the Empire got no nearer than before to its hope of adequate radio communication! The Marconi Company's control of important patents kept other private enterprise from seeking to enter the field on anything like the same scale—the private "competition" which the government had hoped to create was an economic impossibility . . . while at the same time the government platform of competition between its own and privately owned stations was fatal to any agreement between the two as to a division of traffic. . . .

By 1924 Britain had fallen far behind in radio. Once more the situation was handed over to an Imperial Committee—the sixth within twelve years. . . . Spurred by the urgency, it hurried its deliberations and made its report in February.

It went back to the dictum of exclusive government ownership within the dominions!

Meantime Guglielmo Marconi, now a *Senatore* of Italy, thirty years older than the young student of the University of Bologna, had been pursuing his original dream. He had seen the radio to which he had given birth grow past and beyond him, had seen other men take it up and do with it that which he himself had been unable to achieve. Langmuir,

Arnold, the laboratories had made of it an exact science; it was now a thing of vacuum tubes, electrons . . . broadcasting had swept into the world and the whole art had taken on a new character.

And yet Marconi clung to his original idea—telegraphy without wires from point to point. . . .

The whole growth of transoceanic radio had been one of size and power. The seas had been bridged when man had learned to make stations large enough to crash great long waves across them. But lately a new field had opened. The short waves, down below 100 meters, which had long been considered poor, inefficient, because of the irregularity with which signals carried by them had been received, had been revealing themselves capable of unsuspected achievements.

Experiments with these waves had been begun in 1916, following the advent of the tube transmitter which made their generation feasible. Congestion in the upper wave bands, brought about by the growth of broadcasting and of all radio activity, had stimulated this experimentation. By 1922 the laboratories and the more advanced amateurs had been making tests with surprising results. Using very low powers they had been able to communicate great distances; in the winter of that year American amateurs sent messages to England using only a fraction of the power necessary for long waves. . . . Short wave communication, however, was highly subject at that time to atmospheric interference; results by night were far better than by day . . . the problem of increasing power to overcome these conditions was difficult because of the limitation in size placed upon the apparatus by the high radio frequencies necessary. . . .

Anything that would increase distance attracted Marconi, and in those years he was hard at work, investigating the new phenomena, taking long cruises in his yacht *Elettra*.

To increase the strength of the signals, he proposed to go back to something he had tried in his earliest experiments (that, indeed, Hertz himself had tried and accomplished)—he proposed to focus them in one direction by placing an electrical reflector behind his antenna . . . he called this focused signal a radio "beam." At Poldhu, site of the station which had sent his 1901 trans-Atlantic "S," he erected a new short-wave "beam" experimental station in 1923. He heard signals sent from it on 97 meters at 12 kilowatts power 1250 miles by day, 2320 miles by night . . . static was reduced far below that heard on long waves.

Stimulated by this success, he kept on experimenting. In April and May, 1924, he carried out successful tests with Sydney, Australia, 12,000 miles away . . . and on May 30th communicated with the same city by radiotelephony. Using fifteen kilowatts of power, the new station was able to send messages to Canada, the United States, Brazil,

Argentina, India, and South Africa—on eighty-seven meters.

That month the British Marconi Company announced that "beam" radio was a proven success.

The effect of this announcement upon the recent British Imperial Wireless Committee report was the negation of the high-power ideas contained therein. Indeed, this announcement sounded the death-knell of super-power radio, at least as it had been conceived in the past. In two short months the Committee revised its entire plan; in July it signed a contract with Marconi for beam stations.

This contract was an unusual testimonial to the faith of the British Marconi Company in their new apparatus. It provided for a beam system, to be built at Marconi Company expense, but ultimately to be government owned (through the Post Office). The first stations were to be in England and Canada, and provisions were to be made for extending the English station, later, to service with corresponding stations in South Africa, India, and Australia. Sites for the stations were to be provided by the government; the power for sending was to be at least twenty kilowatts; the beam was to be concentrated within an angle no greater than 30°, and the receiving station was to have a "reflector" to gather in the incoming waves; the stations were to be "duplexed," and capable of remote control (that in England from the Central Telegraph Office in London); they were to be completed within 26 weeks from the time the sites were placed at the Company's disposal; they were to be capable of a speed of 100 five-letter words a minute each way during a daily average of (for the Canada circuit) eighteen hours.

The Marconi Company was to receive no money until the British station had been completed and demonstrated by actual operation for seven days! If it fulfilled the 18-hour—100 word per minute condition during this test, the company was to receive half the cost and the station was to be handed over to the Postmaster General. After a further period of six months' satisfactory operation twenty-five per cent more was to be paid . . . and at the end of the first year, the remaining twenty-five per cent! And if at any time during this period the station did not comply with the requirements (or the Canadian station had not been built) the Postmaster was to be free to reject the English station, in which case the Company must refund the money that had been paid to it! The Company's remuneration, if the stations were successful, was to be cost, plus five per cent for establishment charges, and ten per cent for profit, with a maximum government expenditure for the English station (with its four extensions) of £165,333 (a sum which to its sorrow the Marconi Company exceeded considerably . . .).

A good deal of criticism was leveled at the government for entering into this "experiment" instead of sticking to the old tried and true super-power idea, to which the reply was that the Marconi Company took all the risks.

In any case, the government went ahead with a super-power station to supplement the beam. This station—at Rugby, and one of the most powerful in the world—was finished and opened to service early in 1926; it used 54 huge vacuum tubes to generate its continuous waves, and operated on a wave length of 18,740 meters; it had twelve steel masts 820 feet high, supporting three miles of aereal . . . it had cost £490,000.

In November of that year 1926, the English and Canadian beam stations had passed their official tests and were opened to traffic. The Australian circuit was put into operation in April, 1927; that to South Africa in June, 1927; that to India in September.

Thus, through the new Marconi beam, the old idea of linking the dominions of the British Empire by radio, first proposed in 1911, was at last brought to fruition, and with results far superior than could have been dreamed of in those days when men were only hoping to break down the ocean barriers. For the beam system exceeded its contract obligations in every way. Beam signals penetrated static so well that they were splendidly adapted to high-speed automatic sending and receiving; instead of the contracted 100 words per minute, the Marconi beam stations were able to handle an average of 150 . . . in 1928 it was estimated that thirty-five million words were being exchanged annually over the Australian, Canadian, Indian, and South African circuits, at rates considerably lower than those of the cable companies . . . and these twenty-five kilowatt stations were achieving that for which it had been earlier planned to use seven hundred and fifty kilowatts!

Under the agreement between the Radio Corporation of America and the British Marconi Company whereby they exchanged information and were mutually licensed to use each other's patents, the beam system was made available in the United States as soon as was the case abroad.

The engineers in the laboratories of the Radio Corporation, General Electric, and Westinghouse were able to put their knowledge behind the new idea and bring it to greater perfection than ever, both to the benefit of this country and of nations across the water. Both General Electric and the Radio Corporation developed new and better types of beam transmitters and antennae; they cut the beam angle far below the original 30°; by 1928 beam circuits and short waves were carrying the bulk of America's transoceanic traffic and a revolution as sweeping as that brought about by the development of the Alexanderson Alternator had taken place. Those great humming alternators still formed the basic units of the various Radio Centrals; coupled to their giant antennae they still seemed to dominate . . . but the smaller, seemingly insignificant masts that lay beneath those tall steel towers were now the more important. Within the power houses, mute looking groups of switchboards and instrument panels were the transmitters . . . mute looking until one went behind them and saw the glowing tubes with their silent ionic forces . . . doing the work that had been the province of the Alternator. . . .

The increase in short-wave knowledge made it possible, then, for the amateur to send his voice around the world. It was a commonplace to pick a broadcast program out of space, convert it to a short wave-length and rebroadcast it to the ends of the earth. The explorer, deep in the arctic, could communicate with civilization with less power than would light an electric lamp . . . the airman could speak from midocean. . . .

And it was possible for a man in America to go to his home telephone and send his voice to the telephone of him in Europe who had become his ethereal next-door neighbor. This had been the 1927 achievement of the Bell Laboratories, Western Electric, and the Telephone Company, working in coöperation with the Radio Corporation and in part using its equipment . . . it had been their dream for over ten years. In 1923 they had been able to get fifteen words out of every hundred across to England; in 1924 it was sixty; in 1925 it was ninety . . . finally perfected, the commercial installation had cost five million dollars and the efforts of many men for many years—it was equipped to work on long waves under certain atmospheric conditions, short waves under others, and following the opening of service, when only the cities of New York and London were connected, its scope was extended until almost any telephone in the United States could speak to almost any telephone in Europe. . . .

Apart from all this elaborate and varied trans-oceanic communication carried on by private enterprise, there was still the United States Naval Communications Service . . . quite recovered from the post-war depression, its morale restored, its apparatus abreast the era.

Its basic form remained the same; it was a government owned and operated radio link between the American continent and the nation's dependencies and men-o'-war. Its purpose was defensive and strategic; always ready, it existed to serve when called on, and in the interim to be of service not only to the Navy, but to the maritime and general interests of its owners, the nation's people.

As always, it operated its chain of transoceanic stations. Arlington and Annapolis formed a circuit with Guantanamo, Porto Rico, the Canal Zone, San Diego and Mare Island. Mare Island formed one with Pearl Harbor, Hawaii; Pearl Harbor one with Manila, Guam and Samoa; and Puget Sound one with Cordova and St. Paul, Alaska. . . . The old

Arc installations that had been the service's pride were still the mainstays of these circuits, having been remodeled to reduce the harmonic emissions that had made them use overwide channels in the ether, but they were recognizedly obsolete—so quickly had radio changed. Beside them modern short-wave transmitters and receivers were being set up . . . and making possible the one-time dream of the Communications Service—to be able to get messages to American men-o'-war wherever on the seven seas they might be.

For coastal work 11 medium power and 79 low power stations were in operation; and the initial venture into radio compass stations had been extended until there were, in 1928, fifty-six of them rendering most valuable service, gratis, to all mariners by fixing their positions as they approached American harbors or made their way past dangerous points along the coast.

The Department of Commerce, too, had entered the radio field for safety of life at sea, and operated radio beacons on all coasts at the points where they would be of most benefit to mariners. These, listed in marine publications as were lighthouses and buoys, automatically sent out a characteristic signal on a specified wave length during foggy weather, and at certain times of the day and night irrespective of weather conditions. They could be identified by their signal, and vessels equipped with radio compasses could take bearings on them.

As aviation became developed to the point of regular commercial usage, with air-mail lines making scheduled trips day and night, winter and summer, there grew a need for radio beacons to guide these planes along their routes and to their landing fields when the weather was thick. These were first installed by the Department of Agriculture, later taken over by the Department of Commerce . . . ultimately a simplified form of transmitting and receiving apparatus was developed in the Department's laboratory—the receiver consisted of an instrument mounted among the others before the pilot, which told him visually whether or not he was on his course.

Marine Service for American and other merchantmen was built up by the Radio Corporation of America in addition to its transoceanic work. Besides the marine activities of this company, there existed after the war several others engaged in the ship-shore business, notably the Independent Wireless Telegraph Company and the Ship Owner's Radio Service Company. The first two mentioned maintained shore stations for communication with their ship-rental installations; the latter confined its activities to the ship end, and used the shore facilities of the others. The United Fruit Company continued to maintain its own radio service as always.

Vacuum tube apparatus was made available to the Independent Wireless Telegraph Company and the Ship Owner's Radio Service Company, and to any other competing marine-service companies that might apply for it, when the Radio Corporation announced that in the interests of better communication between ship and shore they would license competing ship-service companies to use tube apparatus and circuits on a royalty basis.

However, as time went on it became evident that the United States shipping activity which had so flourished during the war and in the years immediately following, was not to continue. . . . Under these circumstances the demand for radio service fell off; that which persisted was increasingly for better equipment. The Independent Wireless Telegraph Company, which had stuck to "spark" apparatus, was unable to render the service required, and eventually, on January 1st, 1928, was sold to the Radio Corporation. The marine services of the two companies, then, were merged in a new Radio Corporation subsidiary called the Radiomarine Corporation of America, and as rapidly as possible all ships taking its service were equipped with apparatus that rendered proper communication through the crowded ether channels.

CHAPTER TWELVE

THE ELECTRIC WORD GOES ONWARD

WHEN, in years to come, radio is an old, old story, men will look back and wonder why its popular adoption during this third decade of the twentieth century should have been in so many ways such a confused business. By that time all the multifarious legal problems introduced by the rushing spread of the art will have been adjudicated; all the basic inventions will have been assigned to those due them (in 1928 the "feed-back" or "tube oscillator" patent which had seemed so certainly Armstrong's in 1920 was still undecided—the Supreme Court of the United States was considering whether that 1913 invention was the work of Armstrong or of De Forest!); broadcasting, communications, and radio-manufacture will have become three distinct fields of endeavor; science and economics will have settled the uses to which the space-channels should be put—and politics will perhaps not be so wary of those two factors of human life as it was in 1927 and 1928.

The law passed by Congress in 1927 to regulate radio communications will seem, in 1950, like a most primitive piece of radio legislation. That law attempted, in one pronunciamento, to cover the whole of a field so great that on the one hand it reached into twelve million American homes and on the other carried America's communications to the ends of the earth. It attempted to regulate an art that as it then stood had been literally non-existent six years earlier, and that by virtue of its revolutionary changes during that period gave every indication that six years hence it would have been again reborn. It was basically a law concerned with broadcasting, in which portion of the art's utilization the major difficulties then existed; point to point communications were lumped in—later to suffer from application to them of sections wholly included for broadcast regulating reasons. It voiced not once but repeatedly the current Congressional dread of the radio monopoly bugaboo—at that time being artfully stimulated by lobbyists working in the interests of the non-patent-owning "manufacturers." Other fears found expression in it,—that "vested rights" in the ether would be claimed by those who had hitherto used its channels; that the law's provisions would be so rigid that they would not take care of change; that they would be so lax that they would not regulate. . . . It prohibited "vested rights," foreign ownership of radio stations, sending out paid advertising or other matter without announcing that it was paid for and by whom (so that paid political speeches or propaganda should, theoretically, be branded) . . . it laid down as principle the fact that the "ether" belonged to the whole people . . . that whoever was licensed to use it was but granted a temporary boon. . . . And having hemmed the whole matter around with these "prohibitive" declarations, it handed the actual and difficult job of regulation and administration of radio over to a lay commission of five.

This Radio Commission had been proposed originally as a semi-judicial body of nine, which was to act as a check upon the administration of radio by the Secretary of Commerce (for it was that officer to whom the first drafts of the bill proposed—and rightly—to assign the administrative task ahead) as a means of insuring justice to those regulated, and as a means of insuring that the users of radio did not abuse it. The administrative task was purely a matter of juggling science and economics in the interests of the whole people; the regulatory one was legal. . . . The two must, of course, work hand in hand . . . and for the moment Congress wedded them, passed the Secretary by—and with him, unified control—and made the Commission itself both regulator and administrator—*for one year;* then the Commission was to assume its advisory-judicial capacity . . . it would always be the only body with the power to refuse a station license. . . .

The conception was that during this one year these five men (none of whom was to be "financially interested in the manufacture or sale of radio apparatus or in the transmission or operation of radio telegraphy, radio telephony or radio broadcasting," and no more than three of whom were to be members of the same political party)—that these five men would straighten out the broadcasting tangle, would distribute licenses, wave lengths, power, and periods of time for operation, among the different States and Communities so as to give *fair, efficient and equitable radio service to each of the same,* and would allot these licenses, etc., to those applicants by whom it had determined that the "*public interest, convenience or necessity* would be served" . . . . In short, that these five ordinary men would miraculously achieve that to which no man then living, however experienced, could point the way—an ideal solution of what the country should have in broadcasting and radio-communications.

There were, on March 15th, 1928, when the new Commission held its first meeting, some seven hundred broadcasting stations operating on the eighty-nine frequencies available for broadcast purposes in this country. There were some of them on every wave length and some on the wave lengths in between; their power ranged from fifty watts to fifty thousand; they were scattered from coast to coast, distributed not in accordance with area but with concentration of population—the cities were overserved; in Chicago and New York the stations were drowning each other out—regionally the middle western states had the greatest number. Due to their uncontrolled activities, they had destroyed a very large portion of the theoretical radio audience by their mutual interference!

The Commission appointed was, on the whole, as wise a choice as could have been named under political aegis. It was headed by Admiral Bullard (appointed chairman by President Coolidge), and contained a lawyer, a publicist with some experience in broadcasting, a former Department of Commerce Radio Supervisor, and an author and educator, likewise with some experience in broadcasting. These men met in Washington, and for four days held public hearings, which they invited the radio industry and anyone else interested to attend for the presentation of recommendations and opinions. Apart from this, they had available for self-educational purposes the past learning of Secretary Hoover and the Radio Division of his Department.

After a brief period of intensive study, they arrived at several conclusions. First, that if interference was to be avoided the stations must be got back to the ten kilocycle separations inaugurated by the Second Radio Conference, with fifty kilocycle separations for stations in the same vicinity. Second, that the frequencies allocated to Canada must be cleared at once. Third, that stations within city residential districts, which had increased their power so that they now "blanketed" the area about them, must be made to move to the outskirts of the city if they wished to continue operation. Fourth, that it was the Commission's duty to license temporarily every station already licensed by the Department of Commerce, and to see if it was possible to assign the bands of frequency so that all these 739 stations could continue to exist and render service. . . .

It was the latter provision that made all the difficulty, because, as a matter of fact, it was impossible to carry it out to the satisfaction of the broadcaster, in whose interest it had been promulgated. . . . That fact the Commission did not know definitely at the time. The basic reason for the adoption of the provision was the fear that if any station then existing were arbitrarily ruled from the air, that station would bring suit and secure an injunction that would tie the Commission's hands . . . prolonging the unregulated era. The perplexing question—had the Congress acted in accordance with the constitution in declaring that no broadcaster had vested rights in the ether—had never been adjudicated. That he could not own the ether there was no doubt, but the point was that if the broadcaster, having erected a station that had been legally licensed and operated, were denied the right to continue its use, did or did not that constitute virtual confiscation by rendering his station useless for the purpose for which it had been built? —and could his station be "confiscated" by the government without compensation? . . . The law made no provision for compensation. . . .

Even apart from that there was a genuine query on the part of the Commission as to whether or not developments would demonstrate that the existing number of stations, or even more, could be kept on the air. . . .

There were, it must be understood, two extremes under which the 89 wave lengths could theoretically be allocated to broadcast traffic. As one extreme, there might be but one station placed on each wave length—in which case there would be no direct interference anywhere in the country so long as each station stayed on its own channel; every station could theoretically increase its power until it could be heard at any point in the country, or as a theoretical alternative could erect synchronized sister-stations connected by wire or wireless, utilizing the same wave length. . . . This theory of allocation, developed to its utmost, would mean that every listener in the country would be able to hear any one of 89 different broadcast programs at any one time. . . .

As the other extreme, there was the "set-up" whereby as many as possible stations of very limited power might be crowded onto each frequency. Assuming that fifty 100-watt stations might be placed on each of the 89, this would cover the country with 4,550 stations, each of which might be heard consistently within a radius of about a mile and a half from its antenna. Under this theory of allocation, most of the radio listeners in the United States would be able to hear one station—the rest of the "dial" would bring in the squeals and howls of heterodyning. . . .

The first of these hypothetical allocations had been referred to as an "ideal set-up," but it was not, for in the first place, in the 1927-1928 state of the radio art's development, it was scientifically impossible of achievement since it would demand a power and perfection of both transmitting and receiving apparatus that had not yet been attained. And in the 1927-1928 or any other state of America's economics it was an anachronism. Assuming it possible to construct single stations capable of covering the entire country consistently (which it was not), and that each of these stations could reach seventy-five million adult listeners . . . they would not have flourished any more than eighty-nine *Saturday Evening Post's* could have flourished, because only one of them could be listened to at any one time. The reason why there only three national weekly magazines was the fact that more than that number could not be read through during one week by the average person in his "reading" time. In their field, three of them covered the country completely. Just so, the three or four of those eighty-nine broadcast stations that were best managed and that attracted the national advertisers would be successful; their program quality would draw ahead of that of their unsupported competitors, until by sheer economics the eighty-nine stations would have decreased to eighty-eight, to eighty-seven . . . until there were only a handful left.

If a broadcasting station is to continue broadcasting, it must have listeners!

The Federal Radio Commission had 739 stations on its hands, and in allocating broadcast frequencies to them it took a course that was a compromise between the two extremes mentioned above.

First, it "cleared" twenty-five frequencies and assigned them to the twenty-five broadcasters scattered over the country, which had, in its opinion, best served "public convenience, interest or necessity." These were, more or less axiomatically, the stations which had the largest investments, the best equipments, the highest powers, were the most progressive and best capable of serving the greatest audience. . . .

Next, it set out to juggle the remaining stations about on the remaining sixty-four frequencies as best it could to eliminate interference, seeking to reach a condition in which the audible field of every station was unmolested and whatever heterodyning there was, was confined to that no-man's land beyond their range—where, however, it remained in the ether, an irritant to the listener, though it did not—again theoretically—conflict with the programs of the stations which he was able to "bring in."

The carrying out of this allocation problem would have been simpler had the Commission been able to pick stations up like chessmen and put them into locations scientifically better than those in which they then existed—those in which the concentration of population was greatest. . . . But there the stations were, and if they were all to remain without interference, resort must be had to "division of time"—a thoroughly uneconomic procedure, for it meant the maintenance of two, three, or sometimes four stations and their staffs to serve one frequency in one locality. . . .

An experimental "set-up" following all these ideas was worked out and put into effect on June 15th, 1927. The Commission asked the public not to consider this final, or that it represented the Commission's idea of perfect broadcasting.

The immediate result was, of course, a great diminution of the hitherto existing confusion and interference—the simplest form of control would have insured that to some extent, and this set-up really had considerable merit. It was impossible to tell during the ensuing summer just what degree of perfection had been achieved, but wherever interference was reported the matter was studied, and if feasible, changes were made to eliminate it. The Commission moved cautiously and slowly, for while it was easy to order a broadcaster to change his wave-length or to divide time, doing such a thing involved considerable expense on his part; the Commission was anxious not to cause this unjustifiably.

The interference question was helped considerably during this period by the growing use of the quartz piezo-crystal to control station frequencies. When pressure or tension is applied to crystals of quartz they have the capacity to develop electricity; conversely, when electric charges are applied to them they "dilate," setting up natural vibrations whose frequencies depend upon the physical characteristics of the crystal . . . the thinner it is ground, the higher will be its frequency. Such a piezo-crystal (Greek *piezō,*—press) can be included in an oscillating circuit in such a way that its physical vibrations will control the electrical oscillation of the circuit within a very small fraction of one per cent. If, then, a station granted a license to broadcast on a certain frequency will use a piezo-crystal to keep it accurately on that frequency, it will stay in its proper ethereal traffic lane, will not interfere with its neighbors, and will be less liable, in turn, to find its program subject to their interference. By the end of 1927 two hundred of the broadcasters were crystal equipped.

The improvement in listening conditions brought about a corresponding improvement in station quality, as it was axiomatic that the greater audience a station had, the better would be its programs. When the air was cleared, allowing the average listener to choose any one of half a dozen stations, quality rose as each station made a bid to hold its audience. The situation as regarded program talent and directors became more acute than ever in the hinterland, and the "program service" companies—the National Broadcasting Company and the newly formed Columbia Broadcasting System — were deluged with requests for admission to their "chains."

To grant all such requests would, of course, defeat the ends in view—as a rule only one station in a given locality or section could be admitted as a subscriber to such service (just as the press associations limited the number of newspapers to which they extended their facilities). The service of one chain, to be ideal from the viewpoint of the operator, the advertiser, the broadcaster, and the listener, should reach every listener in the country on one frequency and one only; this ideal could not, however, be practically attained at the time. . . .

The natural tendency was to admit the stations with the largest audiences and the greatest ranges, since this would guarantee the advertiser the largest permanent national "circulation." But the range of any station varied so with weather conditions, that as cooler weather came in the fall of

1927 there were many intermediate points in the country from which the same chain programs could be heard on two and sometimes three frequencies, particularly since the very progressive factors that had induced the Radio Commission to grant twenty-five stations exclusive frequencies, made those stations the most desirable to the chain service companies . . . and made those stations most desire chain programs. Twenty-three of them were either subscribers to or operators of such service. . . .

There was some grumbling, inland, over this duplication of programs; the Commission became faced with the question as to whether the public's interest would best be served by a listener's hearing a few superb or a number of mediocre ones . . . and announced that if any station could demonstrate that, without taking chain programs, it would render service equal to that of those who did, it would be given a "clear" frequency in preference to chain subscribers. . . . But this was making a perfectly safe gesture; no one came forward with convincing proof. Such a change, in fact, would in all likelihood merely have meant a realignment of the chain stations in the long run, for broadcasting was moving toward a stabilized business state of affairs. It could not be other than a losing proposition unless the broadcaster could have the physical ability to reach a large audience. . . .

There was also, at this time, the suggestion made that during the hours in which chain programs were on the air, the stations broadcasting them operate on the same frequency. But the possibility of such a working out of the matter was an economic myth. In the first place, the delicate synchronism necessary would have involved the use of a double number of interconnecting wires. Those already in use cost about $2,400,000 annually; to double this would have ended chain broadcasting overnight. It would have made the advertising rates so high that the advertiser could not have gained commensurate returns . . . and the rates for "sustaining" programs so high that no broadcaster could have afforded to use them. Such a plan, moreover, would have robbed the subscriber station of its identity; it would have wasted the subscribers' frequencies while they were connected to the chain . . . it was impossible.

Even had the "chain" been a network of permanent stations owned by the chain, synchronized to use one frequency . . . and had this chain operated in competition with a number of unconnected single stations, one or the other would have gone down, since the listeners could not have supported both of them. Probably the loser would have been the single station; his quality would have been sufficiently poorer than that of the major undertaking to make it inevitable that he would lose his audience. . . .

With the coming of winter and the return of good broadcasting conditions, it was evident that the allocations made by the Commission were far from perfect. Since their going into effect on June 15th, they had been observed by the Commissioners, by the Radio Inspectors of the Department of Commerce (a number of automobiles had been equipped with radio receiving and test apparatus for this purpose), and by the general public, which had sent in thousands of letters to Washington, praising the work thus far done and pointing out changes that would be desirable from the listener's standpoint.

There had been minor changes made during the summer to rectify some of the outstanding troubles; on December 1st a more general reallocation was made with a view to helping things for the winter season. The Commission was now quite experienced; it had worked hard since its inauguration and had become a reasonably effective body in spite of the difficulties of administering anything by group action—and as the result of its experience, it came at last to the decision that there were too many stations broadcasting . . . that some of them would have to go.

There had been all along the hope that this station-elimination problem would take care of itself. Few stations—particularly not those lame ducks who were causing the congestion—were making any money; it had been felt that there would inevitably come a time when the weaklings would fall out of their own weight. This had happened in some cases, but there were others who demonstrated a marvelous hanging-on power. The secret was, of course, that these hoped for a court decision, some day, that would make an assigned frequency a salable commodity in spite of the 1927 law's prohibition of vested rights . . . they hoped to remain in existence until the allocations had been stabilized and the three-year licenses permissible under the law had been granted—then they hoped that with no new stations being admitted to the ether, the value of their own would increase phenomenally. A place on the air, once you had got it, was in 1927 a precious possession, largely because if it were lost, it could not be regained, either by its former tenant or anybody else.

With the December 1st "set-up" in operation, however, the Commission felt that it could not now be seriously handicapped by an injunction and court action; it made plans that a reduction in the number of stations would be its next step.

It was evident that the work intended by the Congress to be accomplished by the Commission during the year it was to act as an administrative body, was far from finished. Under the law, allocations, etc., would become the province of the Secretary of Commerce after March 15th, 1928 . . . and since an ideal set-up had not yet been achieved and there yet remained much juggling and shifting to be done, Congress was unwilling that this lapse into the Secretary's hands should take place. At the same time, there was a great deal of Congressional dissatisfaction with what the Commission had achieved; when it came to the point of continuing that body's administrative powers for another year, there were many Congressmen to voice objections. The matter had been complicated by the sudden death, on November 24th, 1927, of Admiral Bullard; another of the original commissioners had died, and one had resigned. . . .

Radio was in politics. It was discussed in long sessions of Congressional committees; it found its way to the floor of the House, the Senate. And the burden of the discontent with the way it had been administered found expression in one sore spot . . . certain parts of the country had better radio service than others, although the law said, in general, that they should be equally served! Congressmen from the South, the Rocky Mountain states, pointed to the East and the Middle West and demanded to know why those sections had been "favored" with a greater number of radio stations than their own. The five radio zones had been formed on a population basis (24,000,000 in each, except No. 5, the Rocky Mountain and Pacific Coast states, which had 9,000,000); their areas varied widely. Radio, of course, was a thing of area; its physical coverage of the country had nothing to do with population. Nevertheless, the Congress demanded to know why it was that some of the zones should have better radio service than others.

It might as well have asked why certain sections

of the country had better transportation facilities than others; why New York City had a greater choice of theatrical entertainment than Independence, Kansas; why some states had "highways" and others had "roads"; why some communities had a superabundance of gas and electric utilities and others had neither. . . .

Transportation, communication—these things flow toward concentration of population. They flow toward buying population. Radio had tended toward those places where its limited "coverage" area would enclose the greatest number of people . . . it could no more be legislated equally over the United States than could subway trains. Radio was business, dollars and cents. Even when conducted by a church it was business—the church had so many dollars to spend for the prosecution of its mission; if it prosecuted that mission by radio it did so because it considered radio a more efficient way of carrying the word of God to those whom it desired to hear that word, than were the more conventional means. . . .

Broadcasting was paid for by its listeners, just as was any other non-endowed communication from stump speaking up, and as was all other "publication." The cost of it was not felt, because advertising meant increased sales, lower prices, and the absorption of the advertising costs in this reduction . . . nevertheless that money came out of the listeners' pockets in the long run, and since they paid for broadcasting, broadcasting could not be legislated away from the buying public without violating the laws of economics; of itself it must always tend toward that place where it could fill its legitimate purpose. When artificially used, as was the tax-supported broadcasting of England, which carried no advertising, it lost its vitality—British programs were seventy-five per cent lectures, a meritorious situation, perhaps . . . but lacking verve.

In Washington, however, they were not treating broadcasting as business, as commerce, as part of the nation's structure of manufacture, buying and selling. Congress chose to think of it as amusement that was being granted to some, denied to others . . . and on March 22nd, 1928, after having talked all winter without doing anything, and having at length let the Radio Commission's administrative powers expire—once more took radio administration out of the hands of the Secretary of Commerce, and extended the active life of the Commission for one year (until March 16th, 1929). Then, having done so much, it took a step in the face of every law of science and economics and made a law of man designed to express its political will.

"It is hereby declared," said the law, "that the people of all the zones . . . are entitled to equality of radio broadcasting service, both of transmission and reception . . .", which was an absolute misconception, inasmuch as they were only entitled to it as they were able to and did pay for it through their mass buying power. It was somewhat similar to saying that a man in Independence, Kansas, was entitled to as wide a choice of motion picture shows as was a man in New York City. If, under economic laws, the population of Independence could support only two or three motion picture theaters, then whatever moral rights that town's citizens might be conceived to have, it would be fantastic to enunciate as principle that they had the same right to a selection of any one of twenty cinema entertainments as had the citizen of New York.

But the law did not view radio in this light. It went on to demand specifically that the licensing authority "as nearly as possible make and maintain an equal allocation of broadcasting licenses, wave lengths or bands of frequency, periods of time for operation, and of station power, to each of the said zones, when and insofar as there are applications therefor; and shall make a fair and equitable allocation of licenses, wave lengths, time for operation, and station power to each of the States, the District of Columbia, the Territories and possessions of the United States within each zone, *according to population. . . .*"

Here was legislation that went around in circles; it placed a dozen economic checks upon the broadcaster—it got the listener nothing. The fallacy of its reasoning could be demonstrated by a comparison of the states of Texas and Massachusetts. Texas had 4,660,000 population; Massachusetts had

3,800,000. Each of them, under the terms of this amendment, was entitled to something over fifteen per cent of the power, wave lengths, etc., of its zone—in short, to an approximately equal amount. But the granting of this equal power, etc., to them could never give them equivalent radio service, for Massachusetts had 479 persons to the square mile, while Texas had but seventeen . . . and the fundamental error of this law was that it arbitrarily assigned equal radio facilities to each of them because their populations were the same.

In so doing it limited radio-broadcasting's chances for an economic working out of its difficulties. For in greater or lesser degree there was this same situation between every pair of states, every pair of zones. . . .

The Commission should have been given powers that were more flexible . . . not held to rigidity. These were not matters that could be settled by legislation. The operator of a 50,000 watt station that made money in Massachusetts might have lost money in Texas. Only time could work such problems out; time, and thought, and investment . . . and business, for business was pure economics. You could not put that station down in Texas by law . . . and the proposal to take it away from Massachusetts because Texas did not have its mate was cutting off a nose to spite its face.

The history of what steps the Federal Radio Commission took in the administration of this amendment to the Radio Law of 1927 cannot yet be written. Three of its members were new to their tasks. It was faced with a problem calling for such powers of analysis and interpretation, for such knowledge of economics and science, that the proper solution was a matter for supermen. Characteristically, its chairmanship had been given to a Presidential appointee who knew absolutely nothing about the art which he was called upon to administer; again the Commission spent its first weeks in self-education, as a lawyer and jurist with no prior knowledge of radio found himself embarked on the determination of what should be the future of radio in the United States.

In 1928 the art in all its applications seemed destined to go through a period of political football.

In 1928 transoceanic and other long-distance radio communications were rapidly being revolutionized by phenomenal developments in the use of high frequencies (short waves) and the beam. Indeed, the radio art had revolutionized all communications. Its competition had brought about the first improvements made in the cable for many years; the old type of cable, exceedingly slow, had been threatened with supersession by its 200-word-per-minute radio cousin . . . until a new type, the "loaded" cable, capable of speeds equally great, had been developed. . . .

The use of short waves, with their great propensities to distance, had made all wave-length allocation an international question. The general radio uses to which the various bands of space should be put had been agreed upon by an International Radiotelegraph Conference which had met in Washington in October and November, 1927, and at which eighty nations had joined in harmonious and often erudite deliberation to produce a Convention that was a masterpiece of international concord and problem solving.

But the channels within the bands—which companies, which users should get them? It was the broadcast allocation business all over again, with the same pleas, the same recriminations, and under the law its administration for the United States was another part of the Federal Radio Commission's task. When, late in 1927 and during the spring of 1928, applications for short-wave assignments piled up and outnumbered the short waves available, that body had to delve into the intricate subject of communications—completely different from broadcasting—in an attempt to be as wise as Jove as it granted its precious privileges to those who clamored for them.

This rush to use short waves had certain boom elements. Many applicants wished to use them for purely personal ends; there were companies with widely separated branches, for example, who were desirous of establishing radio - communication between them as a cheaper way of handling inter-branch traffic. When such as these learned that the Commission held as a general premise that short-wave circuits should be open to all the people, they announced their willingness to accept general traffic between the proposed stations—in effect, to go into the communications business.

But few such requests were granted; the majority of the short waves were allocated to the legitimate radio-communications companies. It was recognized that there was a difference between willingness to accept public messages, and being directly engaged in public service. The handling of communications was a business in itself; communications' natural channels, particularly since they were limited, belonged in the hands of those to whom they were a vital essential and who would extract from them the full measure of use.

As a matter of fact, if properly allocated there was less shortage of channels in the ether; used to their full capacity those space-highways were capable of carrying the world's messages, at least for the present. The important point was that they be not wasted, handed over to those who would use them for but a fraction of the traffic which they were capable of handling.

It was all making the world so small. Nations twelve thousand miles apart were neighbors. . . . It was all tending so to unite them, to lead them toward that coöperation that had been the tendency of community always.

One could tell what international communications was going to be by observing what inter-state communication had become in the United States. Once, the telephone had been confined to the community; there had been no links between the telephones of separated communities. Then there had grown the long-distance lines . . . and the system had expanded until it embraced every telephone within the land in one group of coöperating enterprises. . . .

Once the telegraph had linked individual cities—New York and Boston; Washington and Baltimore. Separate companies had operated each circuit. But as faraway states became neighbors, these separate companies had been joined together, for a telegraph company must be able to deliver a message filed at any one of its offices, to any community in the land. . . .

So with the radio and international communications. No longer was a circuit to one country enough to constitute a business. A radio-communications company must have connections with every nation on earth if it were to render full and adequate service. . . .

Indeed, it was coming to be seen that transoceanic telegraphy, whether by radio or cable, was an enterprise best carried out as one public utility; in England a most significant merger had been announced.

For years there had been a belief that cable and radio were natural competitors. It was coming to be seen that they were, rather, natural allies. The cable was primarily valuable for carrying heavy traffic between seacoast points. It was rigid. A message that went into it at one end could only come out of it at the other . . . if it were destined beyond that point, it must be turned over to some other communications agency and relayed. The radio, on the other hand, could go where it willed—a coast line was no barrier to it; it was flexible—it could fly to any nation.

The belief that these two were competitors had seen no farther than trans-Atlantic or trans-Pacific telegraphy; had not taken into account the fact that communications were more important than the method by which they were physically communicated . . . that no communications company could economically be limited to connecting certain specific countries but must always tend to spread . . . that the mediums which would best serve communications companies were matters of economics . . . that while radio had specific advantages of flexibility lacked by the cable, the cable had specific advantages of its own that the radio lacked. . . .

On March 15th, 1927, then, the British commercial cables, concentrated in the Eastern Telegraph and its associated companies (the Electra House group), and the British commercial radio, concentrated in the British Marconi Company, had formed a jointly-owned holding company to take over their mutual interests and communications facilities . . . provisional upon their negotiating an agreement with the British and Dominion governments. Radio and cable, then, were to be under one management in the service of the public. . . .

The idea hung fire for a time, but about a year later an Imperial Wireless and Cable Conference was called to consider the subject, and also the troublesome matter of the future of the government owned and operated cables and the government owned and operated radio. The British government had not been having an entirely happy time of it with its communications businesses. Entered into to provide better service to the dominions at government expense (in the old military days), they had since been in competition both with each other and with the private companies; the whole idea of having two cable mediums and two radio mediums, all four of them in competition, had proved economically disastrous.

The Commission made its report on July 26th, 1928, and recommended that the Electra House-Marconi merger be allowed to take place (as dating from April 1st, 1928), and that a giant British-controlled communications company be formed under private ownership, to take over (1) all the commercial communications facilities of these two groups of companies, (2) the government cables, and (3) the leases of the Post Office beam radio stations!

This was to be a $150,000,000 company; it was to have a monopoly of British international communications, but a regulated one,—it was to be granted a definite net revenue, beyond which only half its profits were to go to the company and the other half were to be applied to a reduction in rates; questions of policy, etc., were to be in the hands of an Advisory Committee upon which were to be represented all parts of the British Empire.

Here was the most efficient modern external-communications scheme yet presented.

The trend in the United States was in the same direction. External radio communications had been extended to twenty-five foreign countries by the Radio Corporation; in the cable field a new company, the International Telephone and Telegraph Company, had appeared on the scene and was now reaching toward radio to amplify its communications facilities. This company had had a career reflective of the era; it had been begun when two brothers named Behn, planters in Porto Rico, had erected a modern telephone system on that island. Their success in this venture had led to further Caribbean telephone systems; they had next been given a contract to erect a complete modern telephone structure in Spain—a stupendous task; they had bought the American Telephone and Telegraph Company's foreign manufacturing and sales subsidiary, the International Western Electric Company, and changed its name to the International Standard

Electric Corporation—it controlled manufacturing and distributing companies in Europe, Asia, and Australia. In 1927 they had acquired the All America Cables, Inc., a group of American-owned cables between the United States, Central and South America. In 1928 they had acquired the Mackay communications interests,—the Commercial Cable Company (with cables to Europe and under the Pacific), the Postal Telegraph Company (nation-wide telegraphed), and the Mackay Radio Company (formerly the Federal Telegraph Company, and which now also had stations on the East coast, including that at Sayville).

In 1928, then, the International Telephone and Telegraph Company had huge and multifarious communications interests . . . but such radio rights as it had acquired from Mackay were limited. It seemed, then, that there would be a very logical tendency for it to absorb, or merge with, the Radio Corporation of America, and that this joint venture would ultimately become America's great agency for international communications.

For the moment, however, the way to such a logical solution of the entire external communications question was uncompromisingly blocked by a provision of the 1927 Radio Law, which flatly prohibited any company from acquiring both wire and radio communications interests for the purpose of establishing a monopoly. However sincere a sense of service, then, might actuate the desire for such a merger, the parties to it were inhibited from taking the risk . . . for it might all be torn down again if they were adjudged monopolizers.

This mandate, included in the law from the belief that cable and radio should be competitors, and intended to kill two birds with one stone by preventing the mythical and much-feared broadcasting monopoly, was typical of the lack of conception with which that law had been framed, and of the dangers of binding a growing art by restrictive legislation. . . .

It was time, in 1928, that an entirely new conception be gotten, particularly in political circles, of what the radio art was, of what communications were. It was time that a modern conception of the term "monopoly" be gotten, and that a whole herd of bugaboos be put aside once and for all as belonging back in the days of the robber barons of industry. It was time that radio in all its forms be recognized as a public utility—that it be sympathetically regulated, not arbitrarily legislated.

It was time that it be seen that we were living in a new era, that this art, born of the laboratory and brought by the laboratory and modern manufacture to development faster than any art had ever developed before, could not be stopped, could not be stifled. Radio, as always, was on the verge of progress to take man's breath away, progress in television, in facsimile-transmission, in new uses of the electron, the vacuum tube, in "talking" films . . . led by science it was bringing to the American people benefits no people had ever had before . . . and if one thing was tending to stifle it, that one thing was the attempt to bend it to old ideas.

Broadcasting was becoming as important an element in the nation's business as was the weekly magazine. It was making for prosperity. No one needed to worry about controlling its programs, its subject matter, in the interests of the people . . . its programs were the most accurate and sensitive barometer of the people's desire ever yet devised. Where stations were playing jazz it was because their listeners wanted jazz and had said so; where they were playing fine music it was for the same reason. The listener was King, and the broadcaster bent his knee. . . . In 1928 Hoover and Smith campaigned for the presidency and made wide use of broadcasting facilities (for which their parties paid)—and Hoover and Smith had a full national audience because they were such prominent people. . . .

During those same hours, had it been an unknown orator haranguing for votes for the same men, the listeners would have been reduced to zero, regardless of political interest. . . .

Radio broadcasting was the safest, most innocuous form of information dissemination that had ever appeared. . . .

The public would not be bored!

The electric word, to establish its mysterious contact, must be truly electric. . . .

In 1950 all these things will be looked on as belonging to the salad days of the radio as a public utility. They will be things to smile at in that world of air transportation, radio vision, and speed that transcends present imagination.

And in 1950 the radio art will have influenced this whole people for more than thirty years, breaking down their distance barriers, making all the world their neighbor, carrying the electric word from coast to coast and nation to nation . . . promoting understanding, sympathy, peace. . . .

It will have played its part in the development of music as the cinema, the camera, the color plate, the lithograph, have played theirs in the development of the graphic arts . . . it will have endowed the musician, created demand for his effort. . . .

It will have played its part in education, and in business, and in happiness. . . .

In 1928 we were watching it grow.

How could we tell . . . any more than Marconi could tell as he pressed a key, that day, and launched his first crackling spark into the ether . . . what it would be?

HISTORY OF BROADCASTING:
Radio To Television
An Arno Press/New York Times Collection

Archer, Gleason L.
Big Business and Radio. 1939.

Archer, Gleason L.
History of Radio to 1926. 1938.

Arnheim, Rudolf.
Radio. 1936.

Blacklisting: Two Key Documents. 1952–1956.

Cantril, Hadley and Gordon W. Allport.
The Psychology of Radio. 1935.

Codel, Martin, editor.
Radio and Its Future. 1930.

Cooper, Isabella M.
Bibliography on Educational Broadcasting. 1942.

Dinsdale, Alfred.
First Principles of Television. 1932.

Dunlap, Orrin E., Jr.
Marconi: The Man and His Wireless. 1938.

Dunlap, Orrin E., Jr.
The Outlook for Television. 1932.

Fahie, J. J.
A History of Wireless Telegraphy. 1901.

Federal Communications Commission.
Annual Reports of the Federal Communications Commission. 1934/1935–1955.

Federal Radio Commission.
Annual Reports of the Federal Radio Commission. 1927–1933.

Frost, S. E., Jr.
Education's Own Stations. 1937.

Grandin, Thomas.
The Political Use of the Radio. 1939.

Harlow, Alvin.
Old Wires and New Waves. 1936.

Hettinger, Herman S.
A Decade of Radio Advertising. 1933.

Huth, Arno.
Radio Today: The Present State of Broadcasting. 1942.

Jome, Hiram L.
Economics of the Radio Industry. 1925.

Lazarsfeld, Paul F.
Radio and the Printed Page. 1940.

Lumley, Frederick H.
Measurement in Radio. 1934.

Maclaurin, W. Rupert.
Invention and Innovation in the Radio Industry. 1949.

Radio: Selected A.A.P.S.S. Surveys. 1929–1941.

Rose, Cornelia B., Jr.
National Policy for Radio Broadcasting. 1940.

Rothafel, Samuel L. and Raymond Francis Yates.
Broadcasting: Its New Day. 1925.

Schubert, Paul.
The Electric Word: The Rise of Radio. 1928.

Studies in the Control of Radio: Nos. 1–6. 1940–1948.

Summers, Harrison B., editor.
Radio Censorship. 1939.

Summers, Harrison B., editor.
A Thirty-Year History of Programs Carried on National Radio Networks in the United States, 1926–1956. 1958.

Waldrop, Frank C. and Joseph Borkin.
Television: A Struggle for Power. 1938.

White, Llewellyn.
The American Radio. 1947.

World Broadcast Advertising: Four Reports. 1930–1932.